Applied Reliability for Engineers

Applied Reliability for Engineers

B.S. Dhillon
Professor
Department of Mechanical Engineering
University of Ottawa
Ottawa, Canada

CRC Press
Taylor & Francis Group
Boca Raton London New York

CRC Press is an imprint of the
Taylor & Francis Group, an **informa** business

First edition published 2021
by CRC Press
6000 Broken Sound Parkway NW, Suite 300
Boca Raton, FL 33487-2742

and by CRC Press
2 Park Square, Milton Park, Abingdon, Oxon OX14 4RN

© 2021 B.S. Dhillon

CRC Press is an imprint of Taylor & Francis Group, LLC

Library of Congress Cataloging-in-Publication Data
Names: Dhillon, B. S. (Balbir S.), 1947– author.
Title: Applied reliability for engineers / B.S. Dhillon,
Professor, Dept. of Mechanical Engineering,
University of Ottawa, Ottawa, Canada.
Description: First edition. |
Boca Raton, FL: CRC Press/Taylor & Francis Group, LLC, 2021. |
Includes bibliographical references and index.
Identifiers: LCCN 2020047378 (print) | LCCN 2020047379 (ebook) |
ISBN 9780367676315 (hardback) | ISBN 9781003132103 (ebook)
Subjects: LCSH: Reliability (Engineering)
Classification: LCC TA169 .D433 2021 (print) | LCC TA169 (ebook) |
DDC 620/.00452–dc23
LC record available at https://lccn.loc.gov/2020047378
LC ebook record available at https://lccn.loc.gov/2020047379

ISBN: 978-0-367-67631-5 (hbk)
ISBN: 978-0-367-67632-2 (pbk)
ISBN: 978-1-003-13210-3 (ebk)

Dedication

This book is dedicated to the University of Ottawa for providing facilities and excellent environment to achieve my academic goals.

Contents

Preface...xiii
Author Biography xv

Chapter 1 Introduction 1
    1.1   Reliability History...1
    1.2   Need of Reliability in Product Design 2
    1.3   Reliability Application and Specialized Areas......................2
    1.4   Terms and Definitions 3
    1.5   Useful Sources for Obtaining Information on
          Reliability..4
    1.6   Military and Other Reliability Documents 6
    1.7   Scope of the Book..7
    1.8   Problems 8

Chapter 2 Basic Mathematical Concepts 11
    2.1   Introduction...11
    2.2   Arithmetic Mean and Mean Deviation.................................11
          2.2.1   Arithmetic Mean...12
          2.2.2   Mean Deviation 12
    2.3   Boolean Algebra Laws 13
    2.4   Probability Definition and Properties...................................14
    2.5   Useful Definitions 16
          2.5.1   Cumulative Distribution Function 16
          2.5.2   Probability Density Function 17
          2.5.3   Expected Value 18
          2.5.4   Laplace Transform..19
          2.5.5   Laplace Transform: Final-Value Theorem 20
    2.6   Probability Distributions 21
          2.6.1   Binomial Distribution...21
          2.6.2   Exponential Distribution 22
          2.6.3   Rayleigh Distribution 23
          2.6.4   Weibull Distribution 24
          2.6.5   Bathtub Hazard Rate Curve Distribution 25
    2.7   Solving First-Order Differential Equations Using
          Laplace Transforms...25
    2.8   Problems 27

Chapter 3   Reliability Basics..31

          3.1   Introduction...31
          3.2   Bathtub Hazard Rate Curve 31
          3.3   General Reliability-Related Formulas 32
                    3.3.1   Failure (or Probability) Density Function 32
                    3.3.2   Hazard Rate Function.......................................33
                    3.3.3   General Reliability Function 34
                    3.3.4   Mean Time to Failure 35
          3.4   Reliability Networks 36
                    3.4.1   Series Network 36
                    3.4.2   Parallel Network 39
                    3.4.3 k-out-of-m Network 42
                    3.4.4   Standby System 44
                    3.4.5   Bridge Network 46
          3.5   Problems 48

Chapter 4   Reliability Evaluation Methods..51

          4.1   Introduction...51
          4.2   Failure Modes and Effect Analysis (FMEA) 51
          4.3   Fault Tree Analysis (FTA)...52
                    4.3.1   Fault Tree Probability Evaluation..........................55
          4.4   Markov Method...57
          4.5   Network Reduction Approach...61
          4.6   Decomposition Approach..62
          4.7   Delta-Star Method...65
          4.8   Probability Tree Analysis..69
          4.9   Binomial Method 71
          4.10   Problems 73

Chapter 5   Robot Reliability 75

          5.1   Introduction...75
          5.2   Robot Failure Classifications, Causes, and Corrective
                    Measures 75
          5.3   Robot Reliability–Related Survey Results and Robot
                    Effectiveness Dictating Factors..77
          5.4   Robot Reliability Measures...78
                    5.4.1   Robot Reliability 78
                    5.4.2   Mean Time to Robot Failure (MTTRF) 79
                    5.4.3   Robot Hazard Rate 81
                    5.4.4   Mean Time to Robot Problems...............................82
          5.5   Reliability Analysis of Electric and Hydraulic Robots............83
                    5.5.1   Reliability Analysis of the Electric Robot.................83
                    5.5.2   Reliability Analysis of the Hydraulic Robot 87

5.6   Models for Performing Robot Reliability and Maintenance
       Studies..91
       5.6.1   Model I 91
       5.6.2   Model II 91
       5.6.3   Model III...95
5.7   Problems 96

Chapter 6   Computer and Internet Reliability...99

6.1   Introduction..99
6.2   Computer Failure Causes and Issues in Computer
       System Reliability...99
6.3   Computer Failure Classifications, Hardware and Software
       Error Sources, and Computer Reliability Measures 101
6.4   Computer Hardware Reliability versus Software
       Reliability..102
6.5   Fault Masking 102
       6.5.1   Triple Modular Redundancy (TMR) 102
                6.5.1.1   TMR System Maximum Reliability
                           with Perfect Voter......................................104
                6.5.1.2   TMR System with Voter
                           Time-Dependent Reliability and Mean
                           Time to Failure 106
       6.5.2   N-Modular Redundancy (NMR) 107
6.6   Software Reliability Assessment Methods............................108
       6.6.1   Classification I: Software Metrics 108
                6.6.1.1   Design Phase Measure 108
                6.6.1.2   Code and Unit Test Phase Measure...........108
       6.6.2   Classification II: Software Reliability Models 109
                6.6.2.1   Mills Model..110
                6.6.2.2   Musa Model 111
       6.6.3   Classification III: Analytical Methods 113
6.7   Internet Facts, Figures, Failure Examples,
       and Reliability-Associated Observations 113
6.8   Internet Outage Categories and an Approach for
       Automating Fault Detection in Internet Services...................114
6.9   Models for Performing Internet Reliability and
       Availability Analysis 116
       6.9.1   Model I 116
       6.9.2   Model II...118
6.10   Problems 120

Chapter 7   Transportation Systems Failures and Reliability Modeling 123

7.1   Introduction...123
7.2   Defects in Vehicle Parts and Classifications of
       Vehicle Failures...123

7.3   Mechanical Failure-Associated Aviation Accidents 124
7.4   Rail Defects and Weld Failures and Mechanical Failure-
        Associated Delays in Commuter Rail Service 126
7.5   Road and Rail Tanker Failure Modes and Failure
        Consequences.. 128
7.6   Ship-Related Failures and their Causes 129
7.7   Failures in Marine Environments and Microanalysis
        Techniques for Failure Investigation.................................... 129
        7.7.1   Thermomechanical Analysis 130
        7.7.2   Thermogravimetric Analysis 130
        7.7.3   Differential Scanning Calorimetry 130
        7.7.4   Fourier Transform Infrared Spectroscopy 131
7.8   Mathematical Models for Performing Reliability
        Analysis of Transportation Systems..................................... 131
        7.8.1   Model I 131
        7.8.2   Model II.. 134
        7.8.3   Model III... 137
        7.8.4   Model IV 140
7.9   Problems 143

Chapter 8   Power System Reliability 147

8.1   Introduction.. 147
8.2   Power System Reliability-Related Terms and
        Definitions.. 147
8.3   Power System Service Performance Indices........................ 148
        8.3.1   Index I... 148
        8.3.2   Index II 148
        8.3.3   Index III.. 149
        8.3.4   Index IV 149
        8.3.5   Index V 149
        8.3.6   Index VI.. 150
8.4   Loss of Load Probability (LOLP) 150
8.5   Availability Analysis of a Single Power
        Generator Unit 151
        8.5.1   Model I 151
        8.5.2   Model II.. 154
        8.5.3   Model III... 156
8.6   Availability Analysis of Transmission and Associated
        Systems 158
        8.6.1   Model I 159
        8.6.2   Model II.. 161
8.7   Problems 164

Chapter 9   Medical Equipment Reliability 167

9.1   Introduction.. 167
9.2   Medical Equipment Reliability-Related Facts and Figures........... 167

9.3   Medical Devices and Medical Equipment/Devices
       Categories 168
9.4   Methods and Procedures for Improving Reliability
       of Medical Equipment 169
       9.4.1   Parts Count Method 169
       9.4.2   Failure Mode and Effect Analysis (FMEA) 170
       9.4.3   General Approach 170
       9.4.4   Fault Tree Analysis (FTA) 171
       9.4.5   Markov Method 171
9.5   Human Error in Medical Equipment 171
       9.5.1   Important Medical Equipment/Device
                Operator Errors 172
       9.5.2   Medical Devices with High Incidence of
                Human Error 172
9.6   Medical Equipment Maintainability and
       Maintenance 173
       9.6.1   Medical Equipment Maintainability 173
                9.6.1.1   Reasons for Maintainability Principles'
                          Application 173
                9.6.1.2   Maintainability Design Factors 173
                9.6.1.3   Maintainability Measures 173
       9.6.2   Medical Equipment Maintenance 175
                9.6.2.1   Indices 176
                9.6.2.2   Mathematical Models 177
9.7   Sources for Obtaining Medical Equipment
       Reliability-Related Data 178
9.8   Useful Guidelines for Healthcare and Reliability
       Professionals for Improving Medical Equipment
       Reliability 179
9.9   Problems 180

Chapter 10 Mining Equipment Reliability 183

10.1   Introduction 183
10.2   Reasons for Improving Mining Equipment
        Reliability and Factors Impacting Mining System
        Reliability 183
10.3   Useful Reliability-Related Measures for Mining
        Equipment 184
10.4   Open-Pit System Reliability Analysis 186
        10.4.1   Open-Pit Parallel System 186
        10.4.2   Open-Pit Series System 188
10.5   Dump-Truck Tire Reliability and the Factors Affecting
        Their Life 190
10.6   Programmable Electronic Mining System Failures 191
        10.6.1   Systematic Failures 192
        10.6.2   Random Hardware Failures 192

10.7   Designing Reliable Conveyor Belt Systems and
        Methods of Measuring Winder Rope Degradation 192
        10.7.1   Magnetic Nondestructive Testing Method 193
        10.7.2   Visual Inspection Method 193
10.8   Typical Mining Equipment Maintenance Errors and
        Factors Contributing to Maintenance Errors 194
10.9   Useful Engineering Design-Related Improvement
        Guidelines for Reducing Mining Equipment Maintenance
        Errors 195
10.10 Problems 195

Chapter 11 Oil and Gas Industry Equipment Reliability 199

11.1   Introduction 199
11.2   Optical Connector Failures 199
11.3   Mechanical Seals' Failures 200
        11.3.1   Mechanical Seals' Typical Failure Modes
                  and Their Causes 201
11.4   Corrosion-Related Failures 202
        11.4.1   Types of Corrosion or Degradation that
                  Can Cause Failure 202
        11.4.2   Corrosion/Condition Monitoring Methods 203
11.5   Fatigue Damage Initiation Assessment in Oil and
        Gas Steel Pipes 204
11.6   Oil and Gas Pipeline Fault Tree Analysis 204
11.7   Common Cause Failures Defense Approach for
        Oil and Gas Industry Safety Instrumented Systems 208
        11.7.1   Common Cause Failures Defense Approach 208
11.8   Problems 211

Index 213

Preface

Nowadays, engineering systems/products are an important element of world economy and each year billions of dollars are spent to develop, manufacture, operate, and maintain various types of engineering systems/products around the globe. The reliability of these systems/products has become important than ever before because of their increasing complexity, sophistication, and non-specialists users. Global competition and other factors are forcing manufacturers to produce highly reliable engineering systems/products.

It means that there is a definite need for reliability and other professionals to work closely during design and other phases. To achieve this goal, it is essential that they have an understanding of each other's specialty/discipline to a certain degree. At present, to the best of author's knowledge, there is no book that is specifically concerned with applied reliability in a wide range of areas. It means, at present, to gain knowledge of each other's specialties, these professionals must study various books, reports, and articles on each of the topics in question. This approach is time-consuming and rather difficult because of the specialized nature of the material involved.

Thus, the main objective of this book is to combine a wide range of applied reliability topics into a single volume and to treat the material covered in such a manner that the reader requires no previous knowledge to understand it. The sources of most of the material presented are given in the reference section at the end of each chapter. This will be useful to readers if they desire to delve more deeply into a specific area or topic. At appropriate places, the book contains examples along with their solutions, and at the end of each chapter there are numerous problems to test the reader's comprehension in the area.

The book is composed of 11 chapters. Chapter 1 presents various introductory aspects of applied reliability including reliability application and specialized areas; useful sources for obtaining information on reliability; and military and other reliability documents. Chapter 2 reviews mathematical concepts considered useful to understand subsequent chapters. Some of the topics covered in the chapter are arithmetic mean and mean deviation, Boolean algebra laws, probability properties, probability distributions, and useful mathematical definitions. Chapter 3 presents various introductory aspects of reliability.

Chapter 4 presents a number of methods considered useful to analyze engineering systems reliability. These methods are failure modes and effect analysis, fault tree analysis, the Markov method, network reduction approach, decomposition approach, delta–star method, and probability tree analysis. Chapter 5 is devoted to robot reliability. Some of the topics covered in the chapter are robot failure classifications, causes, and corrective measures; robot effectiveness dictating factors; robot reliability measures, reliability analysis of electric and hydraulic robots; and models for performing robot reliability and maintenance studies. Chapter 6 presents various important aspects of computer and internet reliability. Some of the topics covered in the chapter are causes of computer failure and issues in computer system reliability, computer failure classifications and computer reliability measures, computer hardware reliability versus software reliability, fault masking, software reliability assessment methods, internet facts, figures, and failure examples, internet outage categories, and an approach for automating fault detection in internet services.

Chapter 7 is devoted to transportation systems failures and reliability modeling. Some of the topics covered in the chapter are mechanical failure-associated aviation accidents, defects in vehicle parts and classifications of vehicle failures, rail defects and weld failures, road and rail tanker failure modes and consequences, ship-related failures and their causes, and mathematical models for performing reliability analysis of transportation systems. Chapter 8 presents important aspects of power system reliability. Some of the topics covered in the chapter are power system service performance indices, loss of load probability, and availability analysis of transmission and associated systems. Chapter 9 is devoted to medical equipment reliability. Some of the topics covered in the chapter are methods and procedures for improving reliability of medical equipment, human error in medical equipment, medical equipment maintainability and maintenance, and useful guidelines for healthcare and reliability professionals for improving medical equipment reliability.

Chapter 10 presents important aspects of mining equipment reliability. Some of the topics covered in the chapter are reasons for improving mining equipment reliability, useful reliability-related measures for mining equipment, programmable electronic mining system failures, typical mining equipment maintenance errors, and useful engineering design-related improvement guidelines for reducing mining equipment maintenance errors. Finally, Chapter 11 is devoted to oil and gas industry equipment reliability. Some of the topics covered in the chapter are optical connector failures, corrosion-related failures, fatigue damage initiation assessment in oil and gas steel pipes, and oil and gas pipeline fault tree analysis.

This book will be useful to many individuals including design engineers; system engineers; reliability and maintainability professionals; engineering administrators; graduate and senior undergraduate students in the area of engineering; researchers and instructors of reliability and maintainability; and engineers-at-large.

The author is deeply indebted to many individuals including family members, colleagues, friends, and students for their invisible inputs. The invisible contributions of my children are also appreciated. Last, but not the least, I thank my wife, Rosy, my other half and friend, for typing this entire book and for timely help in proofreading.

B. Dhillon
University of Ottawa

Author Biography

B.S. Dhillon is a professor of Engineering Management in the Department of Mechanical Engineering at the University of Ottawa. He has served as a Chairman/ Director of Mechanical Engineering Department/Engineering Management Program for over 10 years at the same institution. Dr. Dhillon is the founder of the probability distribution named *Dhillon Distribution/Law/Model* by statistical researchers in their publications around the world. He has published over 403 [(i.e., 245(70 single authored + 175 co-authored) journal and 158 conference proceedings] articles on reliability engineering, maintainability, safety, engineering management, etc. He is or has been on the editorial boards of 12 international scientific journals. In addition, Dr. Dhillon has written 48 books on various aspects of health care, engineering management, design, reliability, safety, and quality published by Wiley (1981), Van Nostrand (1982), Butterworth (1983), Marcel Dekker (1984), Pergamon (1986), and so on. His books are being used in over 100 countries and many of them are translated into languages such as German, Russian, Chinese, Arabic, and Persian (Iranian).

He has served as General Chairman of two international conferences on reliability and quality control held in Los Angeles and Paris in 1987. Dr. Dhillon has also served as a consultant to various organizations and bodies and has many years of experience in the industrial sector. He has lectured in over 50 countries, including keynote addresses at various international scientific conferences held in North America, Europe, Asia, and Africa. In March 2004, Dr. Dhillon was a distinguished speaker at the Conference/Workshop on Surgical Errors (sponsored by White House Health and Safety Committee and Pentagon), held at the Capitol Hill (One Constitution Avenue, Washington, DC).

Dr. Dhillon attended the University of Wales where he received a BS in electrical and electronic engineering and an MS in mechanical engineering. He received a PhD in industrial engineering from the University of Windsor.

1 Introduction

1.1 RELIABILITY HISTORY

The history of the reliability field goes back to the early years of the 1930s when probability concepts were applied to electric power generation–related problems [1,2]. During World War II, Germans applied the basic reliability concepts for improving reliability of their V1 and V2 rockets. In 1947, Aeronautical Radio, Inc., and Cornell University carried out a reliability study of over 100,000 electronic tubes. In 1950, an ad hoc committee on reliability was formed by the United States Department of Defense and in 1952 the committee was transformed to a permanent body: Advisory Group on the Reliability of Electronic Equipment (AGREE) [3]. Also, in 1952, a nowadays widely used exponential probability distribution in reliability field received a distinct boost after the publication of an article [4], presenting failure data and the results of various goodness-of-fit tests for competing failure distribution.

In 1964, the first time a National Symposium on Reliability and Quality Control was held in the United States, and in the following year, the Institute of Electrical and Electronic Engineers (IEEE) formed an organization called the Reliability and Quality Control Society. In 1956 and 1957, the following three important documents concerning reliability appeared:

- 1956: a book entitled *Reliability Factors for Ground Electronic Equipment* [5],
- 1957: first military reliability specification: MIL-R-25717 (USAF): Reliability Assurance Program for Electronic Equipment [6],
- 1957: AGREE report [7].

In 1962, the first master's degree program in system reliability engineering was started at the Air Force Institute of Technology of the United States Air Force (USAF), Dayton, Ohio. Moreover, ever since the inception of the reliability field many organizations and individuals around the globe have contributed to it, and has branched out into many specialized areas including software reliability, mechanical reliability, and human reliability.

Additional information on reliability history is available in *Design Reliability: Fundamentals and Applications* [8].

1.2   NEED OF RELIABILITY IN PRODUCT DESIGN

In the past, there have been many factors responsible for considering reliability in product design including product complexity, past system failures, the awareness of cost effectiveness, insertion of reliability-related clauses in design specifications, competition, and public demand. The first three of these factors is described below in detail:

- **Product complexity.** In this case, even if we consider the increase in the product complexity with respect to parts alone, there has been a phenomenal growth of some products. For example, a typical Boeing 747 jumbo jet was made up of around 4.5 million parts, including fasteners. Even for relatively simpler products, there has been a significant increase in complexity in regard to parts. For example, in 1935 a farm tractor was made up of 1200 critical parts and in 1990 the number increased to approximately 2900.
- **The past system failures:** In this case, well-publicized system failures such as those presented below may have also contributed to more serious consideration of reliability in product design [9–11]:
  - **Point Pleasant Bridge Disaster:** This bridge was located on the West Virginia/Ohio border and collapsed in 1967. This disaster caused 46 deaths and its basic cause was metal fatigue of a critical eye bar.
  - **Space Shuttle Challenger Disaster:** This debacle took place in 1986, in which all crew members lost their lives. The main reason for the occurrence of this disaster was design defects.
  - **Chernobyl Nuclear Reactor Explosion:** This debacle also took place in 1986, in the former Soviet Union, in which 31 deaths occurred. This disaster was also due to design defects.
- **Awareness of Cost-Effectiveness:** In this case, many studies conducted over the years indicated that the most effective for-profit contribution is the involvement of reliability professionals with product designers. In fact, according to the judgments of some experts, if it costs $1 to rectify a design defect prior to the initial drafting release, the cost would increase to around $10 after the final release, $100 at the prototype stage, $1000 at the pre-production stage, and $10,000 at the production stage.

1.3   RELIABILITY APPLICATION AND SPECIALIZED AREAS

Ever since the inception of the reliability field, the reliability discipline has branched into many applications and specialized areas such as follows:

- **Robot reliability.** This is an emerging new area of the application of basic reliability principles to robot-related problems. Over the years, many publications on the subject have appeared, including two textbooks [12, 13].
- **Power system reliability.** This is a well-developed area and is basically concerned with the application of reliability principles to conventional power system–associated problems. Over the years, many books on the subject have appeared, including a large number of other publications [14].

- **Medical equipment reliability.** This is also an emerging new area of the application of basic reliability principles to medical equipment–associated problems. Over the years, many publications on the subject have appeared, including one textbook [15].
- **Computer and Internet reliability.** This is also an emerging new area of the application of basic reliability principles to computer- and Internet-related problems. Over the years many publications on the topic have appeared, including one textbook [16].
- **Mining equipment reliability.** This is also an emerging new area of the application of basic reliability principles to mining equipment reliability–associated problems. Over the years, many publications on the subject have appeared, including one textbook [17].
- **Transportation systems reliability.** This is also an emerging new area of the application of basic reliability principles to transportation systems reliability–related problems. Over the years, many publications on the topic have appeared, including one textbook [18].
- **Oil and gas industry equipment reliability.** This is also an emerging new area of the application of basic reliability principles to oil and gas equipment reliability–associated problems. Over the years many publications on the topic have appeared, including one textbook [19].
- **Software reliability.** This is a very important emerging area of reliability as the use of computers and other electronic devices is increasing at an alarming rate. Many books have been written on this topic alone and a comprehensive list of publications on the topic may be found in Refs. [20,21].
- **Mechanical reliability.** This is concerned with the reliability of mechanical items. Over the years, many textbooks and other publications on the subject have appeared. An extensive list of publications on the topic is available in Dhillon [22].
- **Human reliability.** This is a very important emerging area of reliability, as in the past, many times systems failed not due to technical faults but due to human errors. The first book on the topic appeared in 1986 [23] and an extensive list of publications on the topic is available in Dhillon [21].
- **Structural reliability.** This is concerned with the reliability of engineering structures, in particular civil engineering. Over the years a large number of publications including books have appeared on the topic [24].

1.4 TERMS AND DEFINITIONS

Many terms and definitions are used in the area of reliability engineering. Some of the commonly used terms and definitions are as follows [25–27]:

- **Reliability.** This is the probability that an item will perform its assigned task satisfactorily for the stated time period when used under the specified conditions.
- **Mean time to failure (i.e., for the exponential distribution).** This is the sum of the operating time of given items divided by the total number of failures.

- **Mission time.** This is the time during which an item is carrying out its specified mission.
- **Failure.** This is the inability of an item to function within the specified guidelines.
- **Useful life.** This is the length of time an item functions within an acceptable level of failure rate.
- **Downtime.** This is the time period during which the item is not in a condition to conduct its stated function.
- **Redundancy.** This is the existence of more than one means for accomplishing a stated function.
- **Human reliability.** This is the probability of completing a task/job successfully by humans at any required stage in the system operation within a stated minimum time limit (if the time requirement is specified).
- **Active redundancy.** This is a type of redundancy when all redundant items are functioning simultaneously.
- **Human error.** This is the failure to perform a given task (or the performance of a forbidden action) that could lead to disruption of scheduled operations or result in damage to equipment/property.
- **Availability.** This is the probability that an item/system is available for application or use when required.
- **Maintainability.** This is the probability that a failed item will be repaired to its satisfactorily operating state.

1.5   USEFUL SOURCES FOR OBTAINING INFORMATION ON RELIABILITY

There are many different sources to obtain reliability-related information. They may be classified under various different categories. Four such categories are as follows:

- **Category I: Agencies**
  - Reliability Analysis Center, Rome Air Development Center (RADC), Griffiss Air Force Base, New York, NY 13441-5700
  - System Reliability Service, Safety and Reliability Directorate, UKAEA, Wigshaw Lane, Culcheth, Warrington, WA3 4NE, UK.
  - Government Industry Data Exchange Program (GIDEP), GIDEP Operations Center, US Department of Navy, Naval Weapons Station, Seal Beach Corona, CA 91720
  - National Technical Information Center (NTIS), 5285 Port Royal Road, Springfield, VA 22151
  - American National Standards Institute (ANSI), 11 W. 42nd St., New York, NY 10036
    - National Aeronautics and Space Administration (NASA), Reliability Information Center, George C Marshall Space Flight Center, Huntsville, AL 35812
    - Defense Technical Information Center, DTIC-FDAC, 8725 John J. Kingman Road, Suite 0944, Fort Belvoir, VA 22060-6218

- Technical Services Department, American Society for Quality Control, 611 W. Wisconsin Ave., PO Box 3005, Milwaukee, WI 53201-3005
- Space Documentation Service, European Space Agency, Via Galileo Galilei, Frascati 00044, Italy
- **Category II: Journals**
  - *IEEE Transactions on Reliability*, jointly published by the Institute of Electrical and Electronic Engineers (IEEE) and the American Society for Quality Control (ASQC) five times a year.
  - *Reliability, Quality, and Safety Engineering*, published by the World Scientific Publishing company four times a year.
  - *Quality and Reliability Engineering*, published by John Wiley and Sons four times a year.
  - *Quality and Reliability Management*, published by MCB University Press several times a year.
  - *Microelectronics and Reliability*, published by Pergamon Press 12 times a year.
  - *Reliability Review*, published by the Reliability Division of ASQC four times a year.
  - *Reliability Engineering and System Safety*, published by Elsevier Science Publishers several times a year.
- **Category III: Conference Proceedings**
  - *Proceedings of the Annual Reliability and Maintainability Symposium* (US)
  - *Proceedings of the ISSAT International Conference on Reliability and Quality in Design* (US)
  - *Proceedings of the Annual Reliability Engineering Conference for the Electric Power Industry* (US)
  - *Proceedings of the European Conference on Safety and Reliability* (Europe)
  - *Proceedings of the International Conference on Reliability, Maintainability, and Safety* (China)
  - *Proceedings of the International Conference on Reliability and Exploitation of Computer Systems* (Poland)
- **Category IV: Books**
  - Shooman, M.L., *Probabilistic Reliability: An Engineering Approach*, McGraw-Hill Book Company, New York, 1968.
  - Pierce, W.H., *Failure-Tolerant Computer Design*, Academic Press, New York, 1965.
  - Wilcox, R.H., *Redundancy Techniques for Computing Systems*, Spartan Books, New York, 1962.
  - Billinton, R., *Power System Reliability Evaluation*, Gordon and Breach New York, 1970.
  - Cox, R.F., Walter, M.H., editors, *Offshore Safety and Reliability, Elsevier Applied Science*, London, 1991.
  - Dhillon, B.S., *Computer System Reliability: Safety and Usability*, CRC Press, Boca Raton, Florida, 2013.
  - Dhillon, B.S., *Mining Equipment Reliability, Maintainability, and Safety*, Springer-Verlag, London, 2008.

- Dhillon, B.S., *Safety and Reliability in the Oil and Gas Industry: A Practical Approach*, CRC Press, Boca Raton, Florida, 2016.
- Dhillon, B.S., *Power System Reliability, Safety, and Management*, Ann Arbor Science Publishers, Ann Arbor, Michigan, 1983.
- Guy, G.B., editor, *Reliability on the Move: Safety and Reliability in Transportation*, Elsevier Applied Science, London, 1989.
- Dhillon, B.S., *Transportation Systems Reliability and Safety*, CRC Press, Boca Raton, Florida, 2011.
- Dhillon, B.S., *Robot System Reliability and Safety: A Modern Approach*, CRC Press, Boca Raton, Florida, 2015.
- Dhilon, B.S., *Medical Device Reliability and Associated Areas*, CRC Press, Boca Raton, Florida, 2000.
- Dhillon, B.S., *Design Reliability: Fundamentals and Applications*, CRC Press, Boca Raton, Florida, 1999.

1.6 MILITARY AND OTHER RELIABILITY DOCUMENTS

Over the years, many military and other documents, directly or indirectly, concerning reliability have been developed. Such documents can serve as a useful tool in performing practically inclined reliability studies and associated tasks. Some of these documents are as follows:

- MIL-STD-721, Definitions of Terms for Reliability and Maintainability, Department of Defense, Washington, DC.
- MIL-HDBK-338, Electronic Reliability Design Handbook, Department of Defense, Washington, DC.
- MIL-STD-785, Reliability Program for Systems and Equipment, Development and Production, Department of Defense, Washington, DC.
- MIL-STD-2155, Failure Reporting, Analysis and Corrective Action System (FRACAS), Department of Defense, Washington, DC.
- MIL-HDBK-217, Reliability Prediction of Electronic Equipment, Department of Defense, Washington, DC.
- MIL-STD-756, Reliability Modeling and Prediction, Department of Defense, Washington, DC.
- MIL-STD-781, Reliability Design, Qualification and Production Acceptance Tests: Exponential Distribution, Department of Defense, Washington, DC.
- SAE ARD 50010, Recommended Reliability, Maintainability, Supportability (RMS) Terms and Parameters, Society of Automotive Engineers (SAE), Warrendale, PA.
- MIL-HDBK-251, Reliability/Design Thermal Applications, Department of Defense, Washington, DC.
- MIL-STD-790, Reliability Assurance Program for Electronics Parts Specifications, Department of Defense, Washington, DC.
- MIL-HDBK-189, Reliability Growth Management, Department of Defense, Washington, DC.
- MIL-STD-2074, Failure Classification for Reliability Testing, Department of Defense, Washington, DC.

- MIL-HDBK-H108, Sampling Procedures and Tables for Life and Reliability Testing (Based on Exponential Distribution), Department of Defense, Washington, DC.
- ANSI/AIAA R-013, Recommended Practice for Software Reliability, American National Standards Institute, New York.
- MIL-STD-1629, Procedures for Performing a Failure Mode, Effects and Criticality Analysis, Department of Defense, Washington, DC.
- MIL-STD-690, Failure Rate Sampling Plans and Procedures, Department of Defense, Washington, DC.
- MIL-STD-52779, Software Quality Assurance Program Requirements, Department of Defense, Washington, DC.
- MIL-STD-472, Maintainability Prediction, Department of Defense, Washington, DC.
- MIL-STD-2167, Defense System Software Development, Department of Defense, Washington, DC.
- MIL-STD-1472, Human Engineering Design Criteria for Military Systems, Equipment and Facilities, Department of Defense, Washington, DC.
- MIL-STD-337, Design to Cost, Department of Defense, Washington, DC.
- MIL-STD-1908, Definitions of Human Factors Terms, Department of Defense, Washington, DC.
- MIL-STD-1556, Government Industry Data Exchange Program (GIDEP), Department of Defense, Washington, DC.

1.7 SCOPE OF THE BOOK

Nowadays, engineering systems are a very important element of world economy, and each year, a vast sum of money is spent for developing, manufacturing, operating, and maintaining various types of engineering systems around the globe. Global competition and other factors are forcing manufacturers to produce highly reliable engineering systems/products. Over the years, a large number of journal and conference proceeding articles, technical reports, etc., on reliability of engineering systems have appeared in the literature. However, to the best of the author's knowledge, there is no book that covers the topic of applied reliability only within its framework. This is a significant impediment to information seekers on this topic, because they have to consult various sources.

Thus, the main objectives of this book are the followings:

(i) To eliminate the need for professionals and others concerned with applied reliability to consult various different and diverse sources in obtaining desired information, and
(ii) To provide up-to-date information on the subject. This book will be useful to many individuals including design engineers, system engineers, and reliability professionals concerned with engineering systems, engineering system administrators, researchers and instructors in the area of engineering systems, engineering graduate and senior undergraduate students, and engineers at large.

1.8 PROBLEMS

1. Write an essay on the history of the reliability field.
2. Discuss the need of reliability in product design.
3. Describe the following three application areas of reliability:
   - Medical equipment reliability
   - Power system reliability
   - Robot reliability
4. Describe the following three special areas of reliability:
   - Software reliability
   - Mechanical reliability
   - Human reliability
5. Define the following three terms:
   - Reliability
   - Availability
   - Maintainability
6. List at least six agencies to obtain reliability-related information.
7. Define the following four terms:
   - Redundancy
   - Active redundancy
   - Human error
   - Useful life
8. List at least seven military documents that can serve as a useful tool in performing practically inclined reliability studies and associated tasks.
9. Compare structural reliability with software reliability.
10. List at least five journals and three conference proceedings for obtaining reliability-related information.

REFERENCES

1. Lyman, W.J., Fundamental consideration in preparing a master system plan, *Electrical World*, Vol. 101, 1933, pp. 778–792.
2. Smith, S.A., Service reliability measured by probabilities of outage, *Electrical World*, Vol. 103, 1934, pp. 371–374.
3. Coppola, A., Reliability engineering of electronic equipment: a historical perspective, *IEEE Transactions on Reliability*, Vol. 33, 1984, pp. 29–35.
4. Davis, D.J., An analysis of some failure data, *Journal of American Statistical Association*, Vol. 47, 1952, pp. 113–150.
5. Henney, K., Ed., *Reliability Factors for Ground Electronic Equipment*, McGraw-Hill, New York, 1956.
6. MIL-R-25717 (USAF), *Reliability Assurance Program for Electronic Equipment*, Department of Defense, Washington, DC.
7. AGREE Report, Advisory Group on Reliability of Electronic Equipment (AGREE), Reliability of Military Electronic Equipment, Office of the Assistant Secretary of Defense (Research and Engineering), Department of Defense, Washington, DC, 1957.
8. Dhillon, B.S., *Design Reliability: Fundamentals and Applications*, CRC Press, Boca Raton, Florida, 1999.

9. Dhillon, B.S., *Advanced Design Concepts for Engineers*, Technomic Publishing Company, Lancaster, PA, 1998.
10. Elsayed, E.A., *Reliability Engineering*, Addison Wesley Longman, Reading, MA, 1996.
11. Dhillon, B.S., *Engineering Design: A Modern Approach*, Richard D. Irwin, Chicago, IL, 1996.
12. Dhillon, B.S., *Robot Reliability and Safety*, Springer-Verlag, New York, 1991.
13. Dhillon, B.S., *Robot System Reliability and Safety*, CRC Press, Boca Raton, Florida, 2015.
14. Dhillon, B.S., *Power System Reliability, Safety and Management*, Ann Arbor Science, Ann Arbor, MI, 1983.
15. Dhillon, B.S., *Medical Device Reliability and Associated Areas*, CRC Press, Boca Raton, Florida, 2000.
16. Dhillon, B.S., *Computer System Reliability: Safety and Usability*, CRC Press, Boca Raton, Florida, 2013.
17. Dhillon, B.S., *Mining Equipment Reliability, Maintainability, and Safety*, Springer-Verlag, London, 2008.
18. Dhillon, B.S., *Transportation Systems Reliability and Safety*, CRC Press, Boca Raton, Florida, 2011.
19. Dhillon, B.S., *Safety and Reliability in the Oil and Gas Industry: A Practical Approach*, CRC Press, Boca Raton, Florida, 2016.
20. Dhillon, B.S., *Reliability in Computer System Design*, Ablex Publishing, Norwood, NJ, 1987.
21. Dhillon, B.S., *Reliability and Quality Control: Bibliography on General Specialized Areas*, Beta Publishers, Gloucester, Ontario, 1992.
22. Dhillon, B.S., *Mechanical Reliability: Theory, Models, and Applications*, American Institute of Aeronautics and Astronautics, Washington, DC, 1988.
23. Dhillon, B.S., *Human Reliability: With Human Factors*, Pergamon Press, New York, 1986.
24. Dhillon, B.S., *Reliability Engineering Applications: Bibliography on Important Application Areas*, Beta Publishers, Gloucester, Ontario, 1992.
25. Naresky, J.J., Reliability definitions, *IEEE Transactions on Reliability*, Vol. 19, 1970, pp. 198–200.
26. MIL-STD-721, *Definitions of Effectiveness Terms for Reliability, Maintainability, and Human Factors and Safety*, Department of Defense, Washington, DC.
27. Omdahl, T.P., Ed., *Reliability, Availability, Maintainability (RAM) Dictionary*, ASQC Quality Press, Milwaukee, WI, 1988.

2 Basic Mathematical Concepts

2.1 INTRODUCTION

Just like in the development of other areas of science and technology, mathematics has played an important role in the development of the reliability field also. Although the origin of the word "mathematics" may be traced back to the ancient Greek word "mathema", which means "science, knowledge, or learning", the history of our current number symbols, sometimes referred to as the "Hindu–Arabic numeral system" in the published literature [1], goes back the third century BC [1]. Among the early evidences of these number symbols' use are notches found on stone columns erected by the Scythian Emperor of India named Asoka, in around 250 BC [1].

The history of probability goes back to the gambling manual written by Girolamo Cardano (1501–1576), in which he considered some interesting issues concerning probability [1,2]. However, Pierre Fermat (1601–1665) and Blaise Pascal (1623–1662) were the first two individuals who independently and correctly solved the problem of dividing the winnings in a game of chance [1,2]. Pierre Fermat also introduced the idea of "differentiation". In modern probability theory, Boolean algebra plays a pivotal role and is named after an English mathematician George Boole (1815–1864), who published, in 1847, a pamphlet titled *The Mathematical Analysis of Logic: Being an Essay towards a Calculus of Deductive Reasoning* [1–3].

Laplace transforms, often used in reliability area for finding solutions to first-order differential equations, were developed by a French mathematician named Pierre-Simon Laplace (1749–1827). Additional information on mathematics and probability history is available in Refs. [1,2].

This chapter presents basic mathematical concepts that will be useful to understand subsequent chapters of this book.

2.2 ARITHMETIC MEAN AND MEAN DEVIATION

A set of given reliability data is useful only if it is analyzed properly. More specifically, there are certain characteristics of the data that are useful for describing the nature of a given data set, thus enabling better decisions related to the data. This section presents two statistical measures considered useful for studying engineering system reliability–related data [4,5].

2.2.1   ARITHMETIC MEAN

Often, the arithmetic mean is simply referred to as mean and is expressed by

$$m = \frac{\sum_{i=1}^{k} x_i}{k} \qquad (2.1)$$

where

  m is the mean value (i.e., arithmetic mean).
  k is the number of data values.
  x_i is the data value i, for $i = 1,2,3, \ldots, k$.

Example 2.1

Assume that the quality control department of a company involved in the manufacture of systems/equipment for use in mines inspected six identical systems/equipment and found 2, 1, 5, 4, 7, and 3 defects in each system/equipment. Calculate the average number of defects per system/equipment (i.e., arithmetic mean).

By substituting the given data values into Equation (2.1), we get

$$m = \frac{2+1+5+4+7+3}{6}$$

$$= 3.6$$

Thus, the average number of defects per system/equipment is 3.6. In other words, the arithmetic mean of the given dataset is 3.6.

2.2.2   MEAN DEVIATION

This is a measure of dispersion whose value indicates the degree to which given data set tends to spread about a mean value. Mean deviation is expressed by

$$MD = \frac{\sum_{i=1}^{k} |D_i - m|}{k} \qquad (2.2)$$

where

MD is the mean deviation.
k is the number of data values.
D_i is the data value i, for $i = 1,2,3, \ldots, k$.
$|D_i - m|$ is the absolute value of the deviation of D_i from m.

Example 2.2

Calculate the mean deviation of the dataset provided in Example 2.1. Using the Example 2.1 dataset and its calculated mean value (i.e., $m = 3.6$ defects per system/equipment) in Equation (2.2), we obtain

$$MD = \frac{|2-3.6|+|1-3.6|+|5-3.6|+|4-3.6|+|7-3.6|+|3-3.6|}{6}$$

$$= \frac{1.6+2.6+1.4+0.4+3.4+0.6}{6}$$

$$= 1.6$$

Thus, the mean deviation of the dataset provided in Example 2.1 is 1.6.

2.3 BOOLEAN ALGEBRA LAWS

Boolean algebra plays an important role in various types of reliability studies and is named after George Boole (1813–1864), an English mathematician. Five of the Boolean algebra laws are as follows [3,6]:

- **Law I: Commutative Law:**

$$Y + X = X + Y \tag{2.3}$$

where

Y is an arbitrary set or event.
X is an arbitrary set or event.
$+$ denotes the union of sets.

$$Y.X = X.Y \tag{2.4}$$

where

Dot (.) denotes the intersection of sets. Note that Equation (2.4) sometimes is written without the dot (e.g., *YX*), but it still conveys the same meaning.

- **Law II: Idempotent Law:**

$$XX = X \tag{2.5}$$

$$X + X = X \tag{2.6}$$

- **Law III: Associative Law:**

$$(XY)Z = X(YZ) \tag{2.7}$$

where

Z is an arbitrary set or event.

$$(X + Y) + Z = X + (Y + Z) \tag{2.8}$$

- **Law IV: Distributive Law:**

$$X(Y + Z) = XY + XZ \tag{2.9}$$

$$(X + Y)(X + Z) = X + YZ \tag{2.10}$$

- **Law V: Absorption Law:**

$$X + (XY) = X \tag{2.11}$$

$$X(X + Y) = X \tag{2.12}$$

2.4   PROBABILITY DEFINITION AND PROPERTIES

Probability is defined as follows [7]:

$$P(Z) = \lim_{n \to \infty} \left[\frac{N}{n} \right] \tag{2.13}$$

where

$P(Z)$ is the probability of occurrence of event *Z*.
N is the number of times event *Z* occurs in the *n* repeated experiments.

Some of the basic probability properties are as follows [7,8]:

• The probability of occurrence of event, say Z, is

$$0 \le P(Z) \le 1 \tag{2.14}$$

• The probability of the occurrence and nonoccurrence of an event, say event Z, is always:

$$P(Z) + P(\bar{Z}) = 1 \tag{2.15}$$

where

$P(\bar{Z})$ is the probability of occurrence of event Z.
$P(\bar{Z})$ is the probability of nonoccurrence of event Z.

• The probability of the union of n independent events is

$$P(Z_1 + Z_2 + \ldots + Z_n) = 1 - \prod_{i=1}^{n}(1 - P(Z_i)) \tag{2.16}$$

where

$P(Z_i)$ is the probability of occurrence of event Z_i, for $i = 1, 2, 3,\ldots, n$.

• The probability of the union of n mutually exclusive events is

$$P(Z_1 + Z_2 + \ldots + Z_n) = \sum_{i=1}^{n} P(Z_i) \tag{2.17}$$

• The probability of an intersection of n independent events is

$$P(Z_1 Z_2 Z_3 \ldots Z_n) = P(Z_1)P(Z_2)P(Z_3)\ldots P(Z_n) \tag{2.18}$$

Example 2.3

Assume that a transportation system has two critical subsystems Z_1 and Z_2. The failure of either subsystem can, directly or indirectly, result in an accident. The failure probability of subsystems Z_1 and Z_2 is 0.08 and 0.05, respectively.

Calculate the probability of occurrence of the transportation system accident if both of these subsystems fail independently.

By substituting the given data values into Equation (2.16), we get

$$P(Z_1 + Z_2) = 1 - \prod_{i=1}^{2}(1 - P(Z_i))$$

$$= P(Z_1) + P(Z_2) - P(Z_1)P(Z_2)$$

$$= 0.08 + 0.05 - (0.08)(0.05)$$

$$= 0.126$$

Thus, the probability of occurrence of the transportation system accident is 0.126.

2.5   USEFUL DEFINITIONS

This section presents a number of mathematical definitions considered useful for performing various types of applied reliability studies.

2.5.1   CUMULATIVE DISTRIBUTION FUNCTION

For continuous random variables, this is expressed by [7]

$$F(t) = \int_{-\infty}^{t} f(y)dy \qquad (2.19)$$

where

  y is a continuous random variable.
  $f(y)$ is the probability density function.
  $F(t)$ is the cumulative distribution function.

  For $t = \infty$, Equation (2.19) yields

$$F(t) = \int_{-\infty}^{\infty} f(y)dy = 1 \qquad (2.20)$$

It simply means that the total area under the probability density function curve is equal to unity.

  It is to be noted that normally in reliability studies, Equation (2.19) is simply written as

$$F(t) = \int_{0}^{t} f(y)dy \qquad (2.21)$$

Example 2.4

Assume that the probability (i.e., failure) density function of a transportation system is

$$f(t) = \lambda e^{-\lambda t}, \text{ for } t \geq 0, \lambda > 0 \qquad (2.22)$$

where

t is a continuous random variable (i.e., time).
λ is the transportation system failure rate.
f(t) is the probability density function (normally, in the area of reliability engineering, it is referred to as the failure density function).

Obtain an expression for the transportation system cumulative distribution function by using Equation (2.21).
  By substituting Equation (2.22) into Equation (2.21), we obtain

$$F(t) = \int_0^t \lambda e^{-\lambda t} dt$$

$$= 1 - e^{-\lambda t} \qquad (2.23)$$

Thus, Equation (2.23) is the expression for the transportation system cumulative distribution function.

2.5.2 PROBABILITY DENSITY FUNCTION

For a continuous random variable, the probability density function is expressed by [7]

$$f(t) = \frac{dF(t)}{dt} \qquad (2.24)$$

where

f(t) is the probability density function.
F(t) is the cumulative distribution function.

Example 2.5

With the aid of Equation (2.23), prove that Equation (2.22) is the probability density function.

By inserting Equation (2.23) into Equation (2.24), we obtain

$$f(t) = \frac{d(1-e^{-\lambda t})}{dt}$$

$$= \lambda e^{-\lambda t} \tag{2.25}$$

Equations (2.22) and (2.25) are identical.

2.5.3 Expected Value

The expected value of a continuous random variable is expressed by [7]

$$E(t) = \int_{-\infty}^{\infty} t f(t) dt \tag{2.26}$$

where

$E(t)$ is the expected value (i.e., mean value) of the continuous random variable t.

Similarly, the expected value, $E(t)$, of a discrete random variable t is expressed by

$$E(t) = \sum_{j=1}^{n} t_j f(t_j) \tag{2.27}$$

where

n is the number of discrete values of the random variable t.

Example 2.6

Find the mean value (i.e., expected value) of the probability (failure) density function defined by Equation (2.22).

By inserting Equation (2.22) into Equation (2.26), we obtain

$$E(t) = \int_0^{\infty} t \lambda e^{-\lambda t} dt$$

$$= \left[te^{-\lambda t} \right]_0^{\infty} - \left[\frac{-e^{-\lambda t}}{\lambda} \right]_0^{\infty}$$

$$= \frac{1}{\lambda} \tag{2.28}$$

Thus, the mean value of the probability (failure) density function expressed by Equation (2.22) is given by Equation (2.28).

2.5.4 LAPLACE TRANSFORM

The Laplace transform (named after a French mathematician, Pierre-Simon Laplace (1749–1827) of a function, say $f(t)$, is defined by [1,9,10].

$$f(s) = \int_0^\infty f(t)e^{-st}\,dt \qquad (2.29)$$

where

t is a variable.
s is the Laplace transform variable.
$f(s)$ is the Laplace transform of function, $f(t)$.

Example 2.7

Obtain the Laplace transform of the following function:

$$f(t) = e^{-\theta t} \qquad (2.30)$$

where

θ is a constant.

By inserting Equation (2.30) into Equation (2.29), we obtain

$$f(s) = \int_0^\infty e^{-\theta t}e^{-st}\,dt$$

$$= \left[-\frac{e^{-(s+\theta)t}}{(s+\theta)} \right]_0^\infty$$

$$= \frac{1}{(s+\theta)} \qquad (2.31)$$

Thus, Equation (2.31) is the Laplace transform of Equation (2.30).

Laplace transforms of some commonly occurring functions used in applied reliability-related analysis studies are presented in Table 2.1 [9,10].

TABLE 2.1
Laplace transforms of some functions.

f(t)	f(s)
t	$\dfrac{1}{s^2}$
t^m, for $m = 0, 1, 2, 3\ldots$.	$\dfrac{m!}{s^{m+1}}$
K (a constant)	$\dfrac{K}{s}$
$e^{-\theta t}$	$\dfrac{1}{(s+\theta)}$
$te^{-\theta t}$	$\dfrac{1}{(s+\theta)^2}$
$tf(t)$	$\dfrac{-df(s)}{ds}$
$\theta_1 f_1(t) + \theta_2 f_2(t)$	$\theta_1 f_1(s) + \theta_2 f_2(s)$
$\dfrac{df(t)}{dt}$	$sf(s) - f(0)$

2.5.5   LAPLACE TRANSFORM: FINAL-VALUE THEOREM

If the following limits exist, then the final-value theorem may be expressed as

$$\lim_{t \to \infty} f(t) = \lim_{s \to 0} sf(s) \qquad (2.32)$$

Example 2.8

Prove by using the following equation that the left-hand side of Equation (2.32) is equal to its right-hand side:

$$f(t) = \frac{\lambda_1}{(\lambda_1 + \lambda_2)} + \frac{\lambda_2}{(\lambda_1 + \lambda_2)} e^{-(\lambda_1 + \lambda_2)t} \qquad (2.33)$$

where

λ_1 and λ_2 are the constants.

By inserting Equation (2.33) into the left-hand side of Equation (2.32), we obtain

$$\lim_{t\to\infty}\left[\frac{\lambda_1}{(\lambda_1+\lambda_2)}+\frac{\lambda_2}{(\lambda_1+\lambda_2)}e^{-(\lambda_1+\lambda_2)t}\right]=\frac{\lambda_1}{(\lambda_1+\lambda_2)} \qquad (2.34)$$

Using Table 2.1, we get the following Laplace transforms of Equation (2.33):

$$f(s)=\frac{\lambda_1}{s(\lambda_1+\lambda_2)}+\frac{\lambda_2}{(\lambda_1+\lambda_2)}\frac{1}{(s+\lambda_1+\lambda_2)} \qquad (2.35)$$

By substituting Equation (2.35) into the right-hand side of Equation (2.32), we obtain:

$$\lim_{s\to0}\left[\frac{s\lambda_1}{s(\lambda_1+\lambda_2)}+\frac{s\lambda_2}{(\lambda_1+\lambda_2)}\cdot\frac{1}{(s+\lambda_1+\lambda_2)}\right]=\frac{\lambda_1}{(\lambda_1+\lambda_2)} \qquad (2.36)$$

The right-hand sides of Equations (2.34) and (2.36) are identical. Thus, it proves that the left-hand side of Equation (2.32) is equal to its right-hand side.

2.6 PROBABILITY DISTRIBUTIONS

This section presents a number of probability distributions considered useful for performing various types of studies in the area of applied reliability [11].

2.6.1 BINOMIAL DISTRIBUTION

This discrete random variable probability distribution is used in circumstances where one is concerned with the probabilities of the outcome, such as the number of occurrences (e.g., failures) in a sequence of, say, n trials. More clearly, each trial has two possible outcomes (e.g., success or failure), but the probability of each trial remains constant or unchanged.

This distribution is also known as the Bernoulli distribution, named after its founder Jakob Bernoulli (1654–1705) [1]. The binomial probability density function, $f(y)$, is defined by

$$f(y)=\binom{n}{i}p^y q^{n-y}, \text{ for } y=0,\ 1,\ 2,\ 3,\dots,n \qquad (2.37)$$

where

$$\binom{n}{i}=\frac{n!}{i!(n-i)!}$$

y is the number of non-occurrences (e.g., failures) in n trials.
p is the single trial probability of occurrence (e.g., success).
q is the single trial probability of non-occurrence (e.g., failure).

The cumulative distribution function is given by

$$F(y) = \sum_{i=0}^{y} \binom{n}{i} p^i q^{n-i} \tag{2.38}$$

where

$F(y)$ is the cumulative distribution function or the probability of y or fewer non-occurrences (e.g., failures) in n trials.

Using Equations (2.27) and (2.37), we get the mean or the expected value of the distribution as

$$E(y) = np \tag{2.39}$$

2.6.2 EXPONENTIAL DISTRIBUTION

This is one of the simplest continuous random variable probability distributions that is widely used in the industrial sector, particularly in performing reliability studies. The probability density function of the distribution is defined by [12]

$$f(t) = \alpha e^{-\alpha t}, \text{ for } \alpha > 0, t \geq 0 \tag{2.40}$$

where

t is the time t (i.e., a continuous random variable).
α is the distribution parameter.
$f(t)$ is the probability density function.

By inserting Equation (2.40) into Equation (2.21), we obtain the following equation for the cumulative distribution function:

$$F(t) = 1 - e^{-\alpha t} \tag{2.41}$$

Using Equations (2.26) and (2.40), we obtain the following expression for the distribution mean value (i.e., expected value):

$$m = E(t) = \frac{1}{\alpha} \tag{2.42}$$

where

  m is the mean value.

Example 2.9

Assume that the mean time to failure of a transportation system is 1500 hours and its times to failure are exponentially distributed. Calculate the transportation system's probability of failure during an 800 hours mission by using Equations (2.41) and (2.42).

  By inserting the specified data value into Equation (2.42), we obtain

$$\alpha = \frac{1}{1500} = 0.0006 \text{ failures per hour}$$

By substituting the calculated and the specified data values into Equation (2.41), we get

$$F(800) = 1 - e^{-(0.0006)(800)}$$

$$= 0.4133$$

Thus, the transportation system's probability of failure during the 800 hours mission is 0.4133.

2.6.3 RAYLEIGH DISTRIBUTION

This continuous random variable probability distribution is named after its founder, John Rayleigh (1842–1919) [1]. The probability density function of the distribution is defined by

$$f(t) = \left(\frac{2}{\mu}\right) t e^{-\left(\frac{t}{\mu}\right)^2}, \quad t \geq 0, \mu > 0 \tag{2.43}$$

where

  μ is the distribution parameter.

By substituting Equation (2.43) into Equation (2.21), we obtain the following equation for the cumulative distribution function:

$$F(t) = 1 - e^{-\left(\frac{t}{\mu}\right)^2} \tag{2.44}$$

By inserting Equation (2.43) into Equation (2.26), we obtain the following equation for the distribution mean value:

$$m = E(t) = \mu\Gamma\left(\frac{3}{2}\right) \tag{2.45}$$

where

$\Gamma(.)$ is the gamma function and is defined by

$$\Gamma(k) = \int_0^\infty t^{k-1} e^{-t}\,dt, \quad \text{for } k > 0 \tag{2.46}$$

2.6.4 WEIBULL DISTRIBUTION

This continuous random variable probability distribution is named after Walliodi Weibull, a Swedish mechanical engineering professor, who developed it in the early 1950s [13]. The distribution probability density function is expressed by

$$f(t) = \frac{bt^{b-1}}{\mu^b} e^{-\left(\frac{t}{\mu}\right)^b}, \text{ for } \mu > 0, b > 0, t \geq 0 \tag{2.47}$$

where

b and μ are the distribution shape and scale parameters, respectively.

By inserting Equation (2.47) into Equation (2.21), we obtain the following equation for the cumulative distribution function:

$$F(t) = 1 - e^{-\left(\frac{t}{\mu}\right)^b} \tag{2.48}$$

By substituting Equation (2.47) into Equation (2.26), we obtain the following equation for the distribution mean value (expected value):

$$m = E(t) = \mu\Gamma\left(1 + \frac{1}{b}\right) \tag{2.49}$$

It is to be noted that exponential and Rayleigh distributions are the special cases of this distribution for $b = 1$ and $b = 2$, respectively.

2.6.5   BATHTUB HAZARD RATE CURVE DISTRIBUTION

The bathtub-shape hazard rate curve is the basis for reliability studies. This continuous random variable probability distribution can represent bathtub-shape, increasing, and decreasing hazard rates.

This distribution was developed in 1981 [14], and in the published literature by other authors around the world, it is generally referred to as the Dhillon distribution/law/model [15–34].

The probability density function of the distribution is expressed by [14]

$$f(t) = b\mu (\mu t)^{b-1} e^{-\left\{ e^{(\mu t)^b} - (\mu t)^b - 1 \right\}}, \ for \ t \geq 0, \mu > 0, b > 0 \qquad (2.50)$$

where

μ and b are the distribution scale and shape parameters, respectively.

By substituting Equation (2.50) into Equation (2.21), we obtain the following equation for the cumulative distribution function:

$$F(t) = 1 - e^{-\left\{ e^{(\mu t)^b} - 1 \right\}} \qquad (2.51)$$

It is to be noted that this probability distribution for $b = 0.5$ gives the bathtub-shaped hazard rate curve, and for $b = 1$ it gives the extreme value probability distribution. More specifically, the extreme value probability distribution is the special case of this probability distribution at $b = 1$.

2.7   SOLVING FIRST-ORDER DIFFERENTIAL EQUATIONS USING LAPLACE TRANSFORMS

Generally, Laplace transforms are used for finding solutions to linear first-order differential equations, particularly when a set or linear first-order differential equations is involved. The example presented below demonstrates the finding of solutions to a set of linear first-order differential equations, describing a transportation system in regard to reliability, using Laplace transforms.

Example 2.10

Assume that a transportation system can be in any of the three states: operating normally, failed due to a hardware failure, or failed due to a human error. The following three first-order linear differential equations describe each of these transportation system states:

$$\frac{dP_0(t)}{dt} + \left(\lambda_h + \lambda_{h\mu} \right) P_0(t) = 0 \qquad (2.52)$$

$$\frac{dP_1(t)}{dt} - \lambda_h P_0(t) = 0 \tag{2.53}$$

$$\frac{dP_2(t)}{dt} - \lambda_{hu} P_0(t) = 0 \tag{2.54}$$

where

$P_i(t)$ is the probability that the transportation system is in state i at time t, for $i = 0$ (operating normally), $i = 1$ (failed due to a hardware failure), and $i = 2$ (failed due to a human error).
λ_h is the transportation system constant hardware failure rate.
λ_{hu} is the transportation system constant human error rate.

At time $t = 0$, $P_0(0) = 1$, $P_1(0) = 0$, and $P_2(0) = 0$.

By using Table 2.1, differential Equations (2.52)–(2.54), and the given initial conditions, we obtain:

$$sP_0(s) - 1 + (\lambda_h + h_u) P_0(s) = 0 \tag{2.55}$$

$$sP_1(s) - \lambda_h P_0(s) = 0 \tag{2.56}$$

$$sP_2(s) - \lambda_{hu} P_0(s) = 0 \tag{2.57}$$

By solving Equations (2.55)–(2.57), we get

$$P_0(s) = \frac{1}{(s + \lambda_h + \lambda_{hu})} \tag{2.58}$$

$$P_1(s) = \frac{\lambda_h}{s(s + \lambda_h + \lambda_{hu})} \tag{2.59}$$

$$P_2(s) = \frac{\lambda_{hu}}{s(s + \lambda_h + \lambda_{hu})} \tag{2.60}$$

By taking the inverse Laplace transforms of Equations (2.58)–(2.60), we obtain

$$P_0(t) = e^{-(\lambda_h + \lambda_{hu})t} \tag{2.61}$$

$$P_1(t) = \frac{\lambda_h}{(\lambda_h + \lambda_{hu})} \left[1 - e^{-(\lambda_h + \lambda_{hu})t} \right] \tag{2.62}$$

$$P_2(t) = \frac{\lambda_{hu}}{(\lambda_h + \lambda_{hu})}\left[1 - e^{-(\lambda_h + \lambda_{hu})t}\right] \qquad (2.63)$$

Thus, Equations (2.61)–(2.63) are the solutions to differential Equations (2.52)–(2.54).

2.8 PROBLEMS

1. Quality control department of a transportation system manufacturing company inspected six identical transportation systems and discovered 5, 2, 8, 4, 7, and 1 defects in each respective transportation system. Calculate the average number of defects (i.e., arithmetic mean) per transportation system.
2. Calculate the mean deviation of the Question 1 dataset.
3. What is idempotent law?
4. Define probability.
5. What are the basic properties of probability?
6. Assume that an engineering system has two critical subsystems X_1 and X_2. The failure of either subsystem can, directly or indirectly, result in an accident. The failure probability of subsystems X_1 and X_2 is 0.06 and 0.11, respectively.

    Calculate the probability of occurrence of the engineering system accident if both of these subsystems fail independently.
7. Define the following terms concerning a continuous random variable:
   - Cumulative distribution function
   - Probability density function
8. Define the following two items:
   - Expected value of a continuous random variable
   - Expected value of a discrete random variable
9. Define the following two items:
   - Laplace transform
   - Laplace transform: final-value theorem
10. Write down the probability density function for the Weibull distribution. What are the special case probability distributions of the Weibull distribution?

REFERENCES

1. Eves, H., *An Introduction to the History of Mathematics*, Holt, Reinhart and Winston, New York, 1976.
2. Owen, D.B., Editor, *On the History of Statistics and Probability*, Marcel Dekker, New York, 1976.
3. Lipschutz, S., *Set Theory*, McGraw Hill, New York, 1964.
4. Speigel, M.R., *Probability and Statistics*, McGraw Hill, New York, 1975.
5. Speigel, M.R., *Statistics*, McGraw Hill, New York, 1961.
6. Report No. NUREG-0492, *Fault Tree Handbook*, U.S. Nuclear Regulatory Commission, Washington, DC, 1981.

7. Mann, N.R., Schefer, R.E., Singpurwalla, N.D., *Methods for Statistical Analysis of Reliability and Life Data*, John Wiley, New York, 1974.

8. Lipschutz, S., *Probability*, McGraw Hill, New York, 1965.

9. Spiegel, M.R., *Laplace Transforms*, McGraw Hill, New York, 1965.

10. Oberhettinger, F., Baddii, L., *Tables of Laplace Transforms*, Springer-Verlag, New York, 1973.

11. Patel, J.K., Kapadia, C.H., Owen, D.H., *Handbook of Statistical Distributions*, Marcel Dekker, New York, 1976.

12. Davis, D.J., Analysis of some failure data, *Journal of the American Statistical Association* 1952, pp. 113–150.

13. Weibull, W., A statistical distribution function of wide applicability, *Journal of Applied. Mechanics*, Vol. 18, 1951, pp. 293–297.

14. Dhillon, B.S., Life distributions, *IEEE Transactions on Reliability*, Vol. 30, 1981, pp. 457–460.

15. Baker, R.D., Non-parametric estimation of the renewal function, *Computers Operations Research*, Vol. 20, No. 2, 1993, pp. 167–178.

16. Cabana, A., Cabana, E.M., Goodness-of-fit to the exponential distribution, focused on Weibull alternatives, *Communications in Statistics-Simulation and Computation*, Vol. 34, 2005, pp. 711–723.

17. Grane, A., Fortiana, J., A directional test of exponentiality based on maximum correlations, *Metrika*, Vol. 73, 2011, pp. 255–274.

18. Henze, N., Meintnis, S.G., Recent and classical tests for exponentiality: a partial review with comparisons, *Metrika*, Vol. 61, 2005, pp. 29–45.

19. Jammalamadaka, S.R., Taufer, E., Testing exponentiality by comparing the empirical distribution function of the normalized spacings with that of the original data, *Journal of Nonparametric Statistics*, Vol. 15, No. 6, 2003, pp. 719–729.

20. Hollander, M., Laird, G., Song, K.S., Non-parametric interference for the proportionality function in the random censorship model, *Journal of Nonparametric Statistics*, Vol. 15, No. 2, 2003, pp. 151–169.

21. Jammalamadaka, S.R., Taufer, E., Use of mean residual life in testing departures from exponentiality, *Journal of Nonparametric Statistics*, Vol. 18, No. 3, 2006, pp. 277–292.

22. Kunitz, H., Pamme, H., The mixed gamma ageing model in life data analysis, *Statistical Papers*, Vol. 34, 1993, pp. 303–318.

23. Kunitz, H., A new class of bathtub-shaped hazard rates and its application in comparison of two test-statistics, *IEEE Transactions on Reliability*, Vol. 38, No. 3, 1989, pp. 351–354.

24. Meintanis, S.G., A class of tests for exponentiality based on a continuum of moment conditions, *Kybernetika*, Vol. 45, No. 6, 2009, pp. 946–959.

25. Morris, K., Szynal, D., Goodness-of-fit tests based on characterizations involving moments of order statistics, *International Journal of Pure and Applied Mathematics*, Vol. 38, No. 1, 2007, pp. 83–121.

26. Na, M.H., Spline hazard rate estimation using censored data, *Journal of KSIAM*, Vol. 3, No. 2, 1999, pp. 99–106.

27. Morris, K., Szynal, D., Some U-statistics in goodness-of-fit tests derived from characterizations via record values, *International Journal of Pure and Applied Mathematics*, Vol. 46, No. 4, 2008, pp. 339–414.

28. Nam, K.H., Park, D.H., Failure rate for Dhillon model, *Proceedings of the Spring Conference of the Korean Statistical Society*, 1997, pp. 114–118.

29. Nimoto, N., Zitikis, R., The Atkinson index, the Moran statistic, and testing exponentiality, *Journal of the Japan Statistical Society*, Vol. 38, No. 2, 2008, pp. 187–205.

30. Nam, K.H., Chang, S.J., Approximation of the renewal function for Hjorth model and Dhillon model, *Journal of the Korean Society for Quality Management*, Vol. 34, No. 1, 2006, pp. 34–39.
31. Noughabi, H.A., Arghami, N.R., Testing exponentiality based on characterizations of the exponential distribution, *Journal of Statistical Computation and Simulation*, Vol. 1, First, 2011, pp. 1–11.
32. Szynal, D., Goodness-of-fit derived from characterizations of continuous distributions, *Stability in Probability, Banach Center Publications*, Vol. 90, Institute of Mathematics, Polish Academy of Sciences, Warszawa, Poland, 2010, pp. 203–223.
33. Szynal, D., Wolynski, W., Goodness-of-fit tests for exponentiality and Rayleigh distribution, *International Journal of Pure and Applied Mathematics*, Vol. 78, No. 5, 2012, pp. 751–772.
34. Nam, K.H., Park, D.H., A study on trend changes for certain parametric families, *Journal of the Korean Society for Quality Management*, Vol. 23, No. 3, 1995, pp. 93–101.

3 Reliability Basics

3.1 INTRODUCTION

The history of the reliability field may be traced back to the early years of 1930s, when probability concepts were applied to problems associated with electric power generation [1–3]. However, the real beginning of the reliability field is generally regarded as World War II, when German scientists applied basic reliability concepts for improving the performance of their V1 and V2 rockets. Today, the reliability field has become a well-developed discipline and has branched out into many specialized areas, including power system reliability, human reliability and error, mechanical reliability, and software reliability [3–5]. There are many reliability basics.

This chapter presents various reliability basics considered useful to understand subsequent chapters of this book.

3.2 BATHTUB HAZARD RATE CURVE

The bathtub hazard rate curve shown in Figure 3.1 is normally used to describe engineering systems' failure rate. As shown in figure the curve is divided into three parts: burn-in period, useful-life period, and wear-out period.

During the burn-in period, the engineering system/item hazard rate (i.e., time-dependent failure rate) decreases with time t, and some of the reasons for to the occurrence of failures during this period are substandard materials and workmanship, poor quality control, poor processes, poor manufacturing methods, human error, and inadequate debugging [6,7]. The other terms used for this region are infant mortality region, break-in region, and debugging region.

During the useful-life period, the hazard rate remains constant and some of the causes for the occurrence of failures in this region are higher random stress than expected, low safety factors, natural failures, abuse, human errors, and undetectable defects. Finally, during the wear-out period, the hazard rate increases with time t and some of the reasons for the occurrence of failures in this region are wear due to aging; wrong overhaul practices; wear due to friction, creep, and corrosion; short designed-in life of the item under consideration; and poor maintenance.

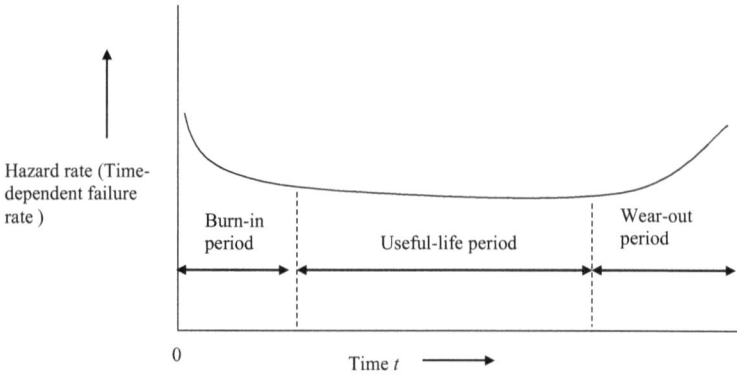

FIGURE 3.1 Bathtub hazard rate curve.

To represent the bathtub hazard rate curve mathematically, shown in Figure 3.1, Equation (3.1) can be used [8]:

$$\lambda(t) = \alpha\beta(\alpha t)^{\beta-1} e^{(\alpha t)^{\beta}} \tag{3.1}$$

where

$\lambda(t)$ is the hazard rate (time-dependent failure rate).
α is the scale parameter.
β is the shape parameter.
t is time.

At $\beta = 0.5$, Equation (3.1) gives the shape of the bathtub hazard rate curve shown in Figure 3.1.

3.3 GENERAL RELIABILITY-RELATED FORMULAS

There are a number of formulas used to perform various types of reliability-related analysis. Four of these formulas are presented in the next four sections.

3.3.1 FAILURE (OR PROBABILITY) DENSITY FUNCTION

The failure (or probability) density function is expressed as shown in Equation (3.2) [7]

$$f(t) = -\frac{dR(t)}{dt} \tag{3.2}$$

where

t is the time.
$f(t)$ is the failure (or probability) density function.
$R(t)$ is the item/system reliability at time t.

Example 3.1

Assume that the reliability of a system is expressed by the following function:

$$R_s(t) = e^{-\lambda_s t} \tag{3.3}$$

where

t is the time.
$R_s(t)$ is the system reliability at time t.
λ_s is the system constant failure rate.

Obtain an expression for the system's failure density function by using Equation (3.2).
    By substituting Equation (3.3) into Equation (3.2), we obtain

$$f(t) = -\frac{de^{-\lambda_s t}}{dt}$$

$$= \lambda_s e^{-\lambda_s t} \tag{3.4}$$

Thus, Equation (3.4) is the expression for the system's failure density function.

3.3.2   HAZARD RATE FUNCTION

This is expressed by

$$\lambda(t) = \frac{f(t)}{R(t)} \tag{3.5}$$

where

$\lambda(t)$ is the item/system hazard rate (i.e., time-dependent failure rate).

By substituting Equation (3.2) into Equation (3.5) we obtain

$$\lambda(t) = -\frac{1}{R(t)} \cdot \frac{dR(t)}{dt} \tag{3.6}$$

Example 3.2

Obtain an expression for the system's hazard rate using Equations (3.3) and
(3.6) and comment on the end result.

By inserting Equation (3.3) into Equation (3.6), we get

$$\lambda(t) = -\frac{1}{e^{-\lambda_s t}} \cdot \frac{de^{-\lambda_s t}}{dt}$$

$$= \lambda_s \tag{3.7}$$

Thus, the system's hazard rate is given by Equation (3.7), and the right-hand
side of this equation is not a function of time t (i.e., independent of time t).
Needless to say, λ_s is normally referred to as the constant failure rate of an item
(in this case, of the system) because it does not depend on time t.

3.3.3 GENERAL RELIABILITY FUNCTION

The general reliability function can be obtained using Equation (3.6). Thus, we have

$$-\lambda(t)dt = \frac{1}{R(t)}.dR(t) \tag{3.8}$$

By integrating both sides of Equation (3.8) over the time interval $[0,t]$, we obtain

$$-\int_0^t \lambda(t)dt = \int_1^{R(t)} \frac{1}{R(t)}dR(t) \tag{3.9}$$

Since, at $t = 0$, $R(t) = 1$.

By evaluating the right-hand side of Equation (3.9) and rearranging, we obtain

$$\ln R(t) = -\int_0^t \lambda(t)dt \tag{3.10}$$

Thus, from Equation (3.10), we obtain

$$R(t) = e^{-\int_0^t \lambda(t)dt} \tag{3.11}$$

Equation (3.11) is the general expression for the reliability function. Thus, it can
be used for obtaining the reliability of an item/system when its times to failure
follow any time-continuous probability distribution (e.g., exponential, Weibull,
and Raleigh).

Example 3.3

Assume that a system's times to failure are exponentially distributed and the constant failure rate is 0.002 failures per hour. Calculate the system's reliability for a 50-hour mission.

By substituting the specified data values into Equation (3.11), we get

$$R(50) = e^{-\int_0^{50} (0.002)dt}$$

$$= e^{-(0.002)(50)}$$

$$= 0.9048$$

Thus, the system's reliability is 0.9048. In other words, there is a 90.48% chance that the system will not malfunction during the stated time period.

3.3.4 MEAN TIME TO FAILURE

The mean time to failure of a system/item can be obtained using any of the following three formulas [7,9]:

$$MTTF = \int_0^\infty R(t)dt \qquad (3.12)$$

or

$$MTTF = E(t) = \int_0^\infty tf(t)dt \qquad (3.13)$$

or

$$MTTF = \lim_{s \to 0} R(s) \qquad (3.14)$$

where

MTTF is the mean time to failure.
E(t) is the expected value.
s is the Laplace transform variable.
R(s) is the Laplace transform of the reliability function *R(t)*.

Example 3.4

Prove by using Equation (3.3) that Equations (3.12) and (3.14) yield the same result for the system's mean time to failure.

By substituting Equation (3.3) into Equation (3.12), we get

$$MTTF_s = \int_0^\infty e^{-\lambda_s t} dt$$

$$= \frac{1}{\lambda_s} \tag{3.15}$$

where

$MTTF_s$ is the system's mean time to failure.

By taking the Laplace transform of Equation (3.3), we obtain

$$R_s(s) = \int_0^\infty e^{-st} e^{-\lambda_s t} dt$$

$$= \frac{1}{s + \lambda_s} \tag{3.16}$$

where

$R_s(s)$ is the Laplace transform of the system reliability function $R_s(t)$.

By inserting Equation (3.16) into Equation (3.14), we get

$$MTTF_s = \lim_{s \to 0} \frac{1}{(s + \lambda_s)}$$

$$= \frac{1}{\lambda_s} \tag{3.17}$$

Equations (3.15) and (3.17) are identical, which proves that Equations (3.12) and (3.14) yield same result for the system's mean time to failure.

3.4 RELIABILITY NETWORKS

An engineering system can form various configurations/networks in conducting reliability analysis. This section is concerned with the reliability evaluation of such commonly occurring networks/configurations.

3.4.1 SERIES NETWORK

This network is the simplest reliability network/configuration, and its block diagram is shown in Figure 3.2.

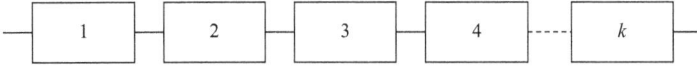

FIGURE 3.2   Block diagram of a k-unit series system (network).

The diagram denotes a k-unit series system (network), and each block in the diagram represents a unit. For the successful operation of the series system, all k units must function normally. In other words, if any one of the k units malfunctions/fails, the series system fails.

The series system, shown in Figure 3.2, reliability is expressed by

$$R_s = P\left(E_1 E_2 E_3 .... E_k\right) \tag{3.18}$$

where

R_s is the series system reliability.
E_i is the successful operation (i.e., success event) of unit i, for $i = 1, 2, 3, ..., k$.
$P\left(E_1 E_2 E_3 .... E_k\right)$ is the occurrence probability of events $E_1, E_2, E_3, ..., E_k$.

For independently failing all units, Equation (3.18) becomes

$$R_s = P\left(E_1\right) P\left(E_2\right) P\left(E_3\right) .... P\left(E_k\right) \tag{3.19}$$

where

$P\left(E_i\right)$ is the probability of occurrence of event E_i, for $i = 1, 2, 3, ..., k$.

If we let $R_i = P\left(E_i\right)$, for $i = 1, 2, 3, ..., k$, Equation (3.19) becomes

$$R_s = R_1 R_2 R_3 .... R_k$$

$$= \prod_{i=1}^{k} R_i \tag{3.20}$$

where

R_i is the unit i reliability for $i = 1, 2, 3, ..., k$.

For constant failure rate λ_i of unit i from Equation (3.11) $\left(\lambda_i\left(t\right) = \lambda_i\right)$, we get

$$R_i\left(t\right) = e^{-\lambda_i t} \tag{3.21}$$

where

$R_i(t)$ is the reliability of unit i at time t.

By substituting Equation (3.21) into Equation (3.20), we get

$$R_s(t) = e^{-\sum_{i=1}^{k} \lambda_i t} \tag{3.22}$$

where

$R_s(t)$ is the series system reliability at time t.

By substituting Equation (3.22) into Equation (3.12), we obtain the following expression for the series system mean time to failure:

$$MTTF_s = \int_0^\infty e^{-\sum_{i=1}^{k} \lambda_i t} \, dt$$

$$= \frac{1}{\sum_{i=1}^{k} \lambda_i} \tag{3.23}$$

where

$MTTF_s$ is the series system mean time to failure.

By substituting Equation (3.22) into Equation (3.6), we obtain the following expression for the series system hazard rate:

$$\lambda_s(t) = -\frac{1}{e^{-\sum_{i=1}^{k} \lambda_i t}} \left[-\sum_{i=1}^{k} \lambda_i \right] e^{-\sum_{i=1}^{k} \lambda_i t}$$

$$= \sum_{i=1}^{k} \lambda_i \tag{3.24}$$

where

$\lambda_s(t)$ is the series system hazard rate.

Here, it is to be noted that the right-hand side of Equation (3.24) is independent of time t. Thus, the left-hand side of this equation is simply λ_s, the failure rate of the series system. It means that whenever we add up the failure rates of items/units, we automatically assume that these items/units form a series network/configuration, a worst-case design scenario in regard to reliability.

Example 3.5

Assume that a system has four independent and identical subsystems, and the constant failure rate of a subsystem is 0.0006 failures per hour. All four subsystems must operate normally for the system to function successfully. Calculate the following:

- System reliability for an eight-hour mission.
- System mean time to failure.
- System failure rate.

In this case, the subsystems of the system form a series configuration/network. Thus, by substituting the given data values into Equation (3.22), we get

$$R_s(8) = e^{-(0.0006)(4)(8)}$$
$$= 0.9809$$

Substituting the given data values into Equation (3.23) yields

$$MTTF_s = \frac{1}{4(0.0006)}$$
$$= 416.66 \text{ hours}$$

Using the specified data values in Equation (3.24) yields

$$\lambda_s = 4(0.0006)$$
$$= 0.0024 \text{ failures per hour}$$

Thus, the system reliability, mean time to failure, and failure rate are 0.9809, 416.66 hours, and 0.0024 failures per hour, respectively.

3.4.2 PARALLEL NETWORK

In this case, the system has k simultaneously operating units/items, and at least one of these units/items must work normally for the successful operation of the system. The k-unit parallel system/network block diagram is shown in Figure 3.3, and each block in the diagram represents a unit.

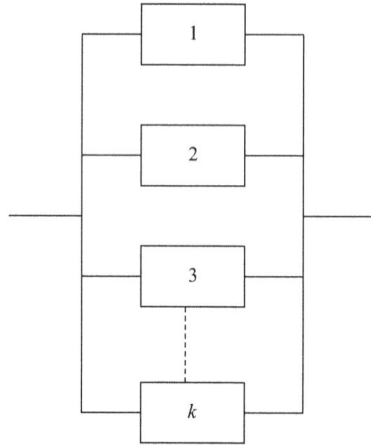

FIGURE 3.3   Block diagram of a parallel system/network with k units.

The failure probability of the parallel system/network shown in Figure 3.3 is expressed by

$$F_p = P\left(\bar{E}_1 \bar{E}_2 \bar{E}_3 .... \bar{E}_k\right) \tag{3.25}$$

where

F_p is the failure probability of the parallel system.
$\bar{E}_i$ is the failure (i.e., failure event) of unit i, for $i = 1, 2, 3,..., k$.
$P(\bar{E}_1 \bar{E}_1 \bar{E}_2 \bar{E}_3 .... \bar{E}_k)$ is the probability of occurrence of events $\bar{E}_1, \bar{E}_2, \bar{E}_3,..., and$
$\bar{E}_k$.

For independently failing parallel units, Equation (3.25) is written as

$$F_p = P(\bar{E}_1)P(\bar{E}_2)P(\bar{E}_3)....P(\bar{E}_k) \tag{3.26}$$

where

$P(\bar{E}_i)$ is the occurrence probability of failure event $\bar{E}_i$, for $i = 1, 2, 3,..., k$.
If we let $F_i = P(\bar{E}_i)$, for $i = 1, 2, 3,..., k$, then Equation (3.26) becomes

$$F_p = F_1 F_2 F_3 .... F_k \tag{3.27}$$

where

F_i is the failure probability of unit i, for $i = 1, 2, 3,..., k$.

By subtracting Equation (3.27) from unity, we obtain

$$R_p = 1 - F_p$$
$$= 1 - F_1 F_2 F_3 .... F_k \qquad (3.28)$$

where

R_p is the reliability of the parallel system/network.

For constant failure rate of λ_i of unit i, subtracting Equation (3.21) from unity and then inserting it into Equation (3.28) yields

$$R_p(t) = 1 - \left(1 - e^{-\lambda_1 t}\right)\left(1 - e^{-\lambda_2 t}\right)\left(1 - e^{-\lambda_3 t}\right)....\left(1 - e^{-\lambda_k t}\right) \qquad (3.29)$$

where

$R_p(t)$ is the parallel system/network reliability at time t.

For identical units, by substituting Equation (3.29) into Equation (3.12), we obtain the following expression for the parallel system/network mean time to failure:

$$MTTF_p = \int_0^\infty \left[1 - \left(1 - e^{-\lambda t}\right)^k\right] dt$$

$$= \frac{1}{\lambda} \sum_{i=1}^{k} \frac{1}{i} \qquad (3.30)$$

where

$MTTF_p$ is the identical units parallel system/network mean time to failure.
λ is the unit constant failure rate.

Example 3.6

Assume that a system has four independent, identical, and active units. At least one of these units must operate normally for the system to operate successfully.
    Calculate the system's reliability if each unit's failure probability is 0.15.
    By inserting the given values into Equation (3.28), we get

$$R_p = 1 - (0.15)(0.15)(0.15)(0.15)$$
$$= 0.9994$$

Thus, the reliability of the system is 0.9994.

Example 3.7

Assume that a system has four independent and identical units in parallel. The constant failure rate of a unit is 0.006 failures per hour. Calculate the system's mean time to failure.

By substituting the given data values into Equation (3.30), we get

$$MTTF_p = \frac{1}{(0.006)}\left[1+\frac{1}{2}+\frac{1}{3}+\frac{1}{4}\right]$$

$$= 347.22 \text{ hours}$$

Thus, the system's mean time to failure is 347.22 hours.

3.4.3 K-OUT-OF-M NETWORK

The k-out-of-m network is another form of redundancy in which at least k units out of a total of m active units must work normally for the successful operation of the system/network. The block diagram of a k-out-of-m unit system/network is shown in Figure 3.4. Each block in the diagram represents a unit.

The parallel and series networks are special cases of this network for $k = 1$ and $k = m$, respectively.

By using the binomial distribution, for identical and independent units, we write down the following expression for reliability of k-out-of-m unit network/system shown in Figure 3.4:

$$R_{k/m} = \sum_{j=k}^{m} \binom{m}{j} R^j (1-R)^{m-j} \tag{3.31}$$

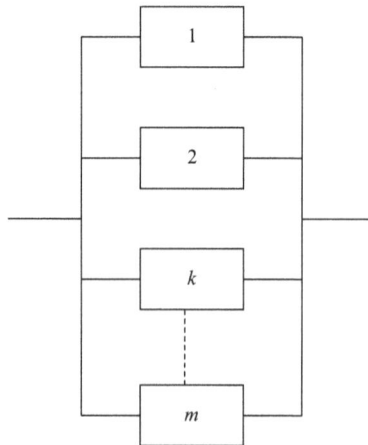

FIGURE 3.4 k-out-of-m unit network/system block diagram.

where

$$\binom{m}{j} = \frac{m!}{(m-j)!i!} \tag{3.32}$$

$R_{k/m}$ is the k-out-of-m network reliability.
R is the unit reliability.

For constant failure rate of the identical units, using Equations (3.11) and (3.31), we obtain

$$R_{k/m}(t) = \sum_{j=k}^{m} \binom{m}{j} e^{-j\lambda t} \left(1 - e^{-\lambda t}\right)^{m-j} \tag{3.33}$$

where

λ is the unit constant failure rate.
$R_{k/m}(t)$ is the k-out-of-m network reliability at time t.

By substituting Equation (3.33) into Equation (3.12), we get

$$MTTF_{k/m} = \int_{0}^{\infty} \left[\sum_{j=k}^{m} \binom{m}{j} e^{-j\lambda t} \left(1 - e^{-\lambda t}\right)^{m-j} \right] dt$$

$$= \frac{1}{\lambda} \sum_{j=k}^{m} \frac{1}{j} \tag{3.34}$$

where

$MTTF_{k/m}$ is the mean time to failure of the k-out-of-m network/system.

Example 3.8

Assume that a system has five active, independent, and identical units in parallel. For the successful operation of the system, at least four units must operate normally. Calculate the mean time to failure of the system if the unit constant failure rate is 0.0005 failures per hour.

By inserting the given data values into Equation (3.34), we obtain

$$MTTF_{4/5} = \frac{1}{(0.0005)} \sum_{j=4}^{5} \frac{1}{j}$$

$$= \frac{1}{(0.0005)} (\frac{1}{4} + \frac{1}{5})$$

$$= 900 \text{ hours}$$

Thus, the mean time to failure of the system is 900 hours.

3.4.4 STANDBY SYSTEM

This is another network or configuration in which only one unit operates and m units are kept in their standby mode. The total system contains $(m + 1)$ units, and as soon as the operating unit fails, the switching mechanism detects the failure and turns on one of the standby units. The system fails when all the standby units fail. The block diagram of a standby system with one functioning and m standby units is shown in Figure 3.5. Each block in the diagram denotes a unit.

Using Figure 3.5 block diagram for independent and identical units, perfect switching mechanism and standby units, and the time-dependent unit failure rate, we write the following expression for the standby system reliability [10]

$$R_{ss}(t) = \sum_{j=0}^{m} \left[\left[\int_0^t \lambda(t) dt \right]^j e^{-\int_0^t \lambda(t) dt} \right] / j! \qquad (3.35)$$

where

$R_{ss}(t)$ is the standby system reliability at time t.
$\lambda(t)$ is the unit time-dependent failure rate/hazard rate.

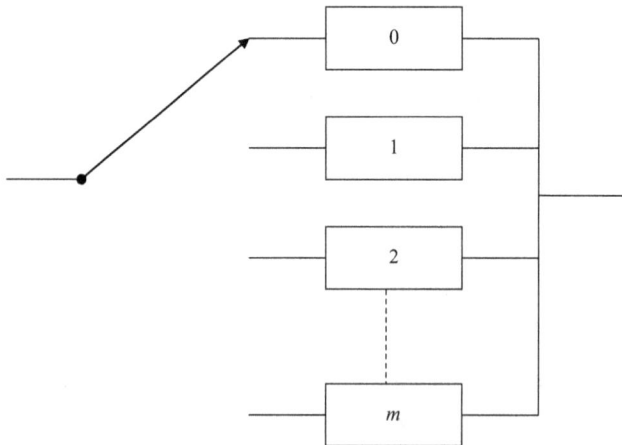

FIGURE 3.5 Block diagram of a standby system containing one operating unit and m standby units.

For unit's constant failure rate (i.e., $\lambda(t) = \lambda$),Equation (3.35) yields

$$R_{ss}(t) = \left[\sum_{j=0}^{m} (\lambda t)^j \right] / j! \qquad (3.36)$$

where

λ is the unit constant failure rate.

By inserting Equation (3.36) into Equation (3.12), we obtain

$$MTTF_{ss} = \int_0^{\infty} \left[\left[\sum_{j=0}^{m} (\lambda t)^j \, e^{-\lambda t} \right] / j! \right] dt$$

$$= \frac{m+1}{\lambda} \qquad (3.37)$$

where

$MTTF_{ss}$ is the standby system mean time to failure.

Example 3.9

Assume that a standby system contains two independent and identical units (i.e., one operating and other on standby). The unit constant failure rate is 0.002 failures per hour.

Calculate the standby system reliability for a 100-hour mission and mean time to failure, assuming that the switching mechanism is perfect and the standby unit remains as good as new in its standby mode.

By inserting the stated data values into Equation (3.36), we get

$$R_{ss}(100) = \sum_{j=0}^{1} \left[\left[(0.002)(100) \right]^j e^{-(0.002)(100)} \right] / j!$$

$$= \frac{\left[(0.002)(100) \right]^0 e^{-(0.002)(100)}}{0!} + \frac{\left[(0.002)(100) \right]^1 e^{-(0.002)(100)}}{1!}$$

$$= 0.8187 + 0.1637$$

$$= 0.9824$$

Similarly, by substituting the given data values into Equation (3.37), we get

$$MTTF_{ss} = \frac{1+1}{(0.002)}$$

$$= 1,000 \text{ hours}$$

Thus, the standby system reliability for a 100-hour mission and mean time to failure are 0.9824 and 1,000 hours, respectively.

3.4.5 Bridge Network

Sometimes units in systems may form a bridge network/configuration, as shown in Figure 3.6. Each block in the figure represents a unit, and all five units are labeled with numerals.

For independently failing units of the bridge network shown in Figure 3.6, reliability is expressed by [11]

$$R_b = 2R_1 R_2 R_3 R_4 R_5 + R_1 R_3 R_5 + R_2 R_3 R_4 + R_2 R_5 + R_1 R_4$$

$$-R_1 R_2 R_3 R_4 - R_1 R_2 R_3 R_5 - R_2 R_3 R_4 R_5 - R_1 R_2 R_4 R_5 - R_3 R_4 R_5 R_1 \qquad (3.38)$$

where

R_b is the reliability of the bridge network.
R_i is the reliability of unit i, for $i = 1, 2, 3, 4, 5$.

For identical units, Equation (3.38) simplifies to

$$R_b = 2R^5 - 5R^4 + 2R^3 + 2R^2 \qquad (3.39)$$

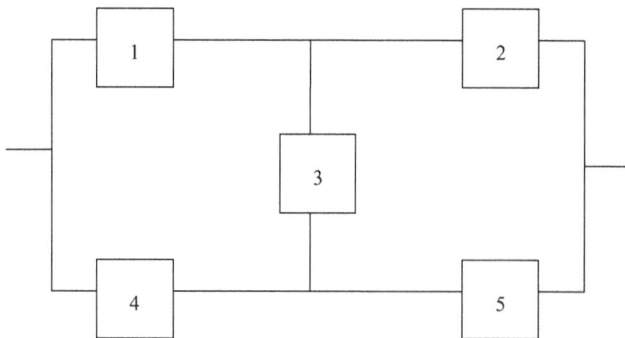

FIGURE 3.6 A five dissimilar unit bridge network/configuration.

where

R is the unit reliability.

For constant unit failure rate using Equation (3.11) and Equation (3.39), we obtain

$$R_b(t) = 2e^{-5\lambda t} - 5e^{-4\lambda t} + 2e^{-3\lambda t} + 2e^{-2\lambda t} \qquad (3.40)$$

where

$R_b(t)$ is the reliability of the bridge network at time t.
λ is the unit constant failure rate.

By inserting Equation (3.40) into Equation (3.12), we obtain

$$MTTF_b = \int_0^\infty \left(2e^{-5\lambda t} - 5e^{-4\lambda t} + 2e^{-3\lambda t} + 2e^{-2\lambda t} \right) dt$$

$$= \frac{49}{60\lambda} \qquad (3.41)$$

where

$MTTF_b$ is the bridge network mean time to failure.

Example 3.10

Assume that a system has five independent and identical units forming a bridge network/configuration and the constant failure rate of each unit is 0.0005 failures per hour.

Calculate the bridge network/configuration reliability for a 250-hour mission and the mean time to failure.

By substituting the given data values into Equation (3.40), we obtain

$$R_b(200) = 2e^{-5(0.0005)(250)} - 5e^{-4(0.0005)(250)} + 2e^{-3(0.0005)(250)} + 2e^{-2(0.0005)(250)}$$

$$= 0.9700$$

Similarly, by inserting the specified data value into Equation (3.41), we get

$$MTTF_b = \frac{49}{60(0.0005)}$$

$$= 1633.33 \text{ hours}$$

Thus, the bridge network/configuration reliability and the mean time to failure are 0.9700 and 1633.33 hours, respectively.

3.5 PROBLEMS

1. Describe the bathtub hazard rate curve and write down the equation that can be used to represent it.
2. Write down the general formulas for the following two functions:
   • Reliability function
   • Hazard rate function
3. Write down three formulas to obtain mean time to failure.
4. Assume that a system has five identical and independent subsystems, and the constant failure rate of a subsystem is 0.0002 failures per hour. All five subsystems must operate normally for the system to operate successfully. Calculate the following:
   • System mean time to failure
   • System reliability for an eighteen-hour mission
   • System failure rate
5. Assume that a system has three identical, independent, and active units. At least one of these units must operate normally for the system to operate successfully. Calculate the following:
   • The system reliability for an eight-hour mission if the constant failure rate of a unit is 0.0002 failures per hour.
6. What are the special case networks of the k-out-of-m network?
7. Assume that a standby system contains three identical and independent units (i.e., one operating, the other two on standby). The unit constant failure rate is 0.008 failures per hour. Calculate the standby system reliability for a 200-hour mission and mean time to failure, assuming that the switching mechanism is perfect and the standby units remain as good as new in their standby modes.
8. Compare the k-out-of-m network with the standby system.
9. Prove Equation (3.34) step-by-step by utilizing Equation (3.33).
10. Assume that a system has five identical and independent units forming a bridge network and the unit constant failure rate is 0.0007 failures per hour. Calculate the system reliability for a 100-hour mission and the mean time to failure.

REFERENCES

1. Layman, W.J., Fundamental consideration in preparing a master plan, *Electrical World*, Vol. 101, 1933, pp. 778–792.
2. Smith, A., Service reliability measured by probabilities of outage, *Electrical World*, Vol. 103, 1934, pp. 371–374.
3. Dhillon, B.S., *Power System Reliability, Safety, and Management*, Ann Arbor Science Publishers, Ann Arbor, Michigan, 1983.
4. Dhillon, B.S., *Human Reliability: With Human Factors*, Pergamon Press, New York, 1986.

5. Dhillon, B.S., *Mechanical Reliability: Theory, Models, and Applications*, American Institute of Aeronautics and Astronautics, Washington, DC, 1988.
6. Kapur, K.C., *Reliability and Maintainability, in Handbook of Industrial Engineering*, edited by G. Salvendy, John Wiley, New York, 1982, pp. 8.5.1–8.5.34.
7. Dhillon, B.S., *Design Reliability: Fundamental and Applications*, CRC Press, Boca Raton, Florida, 1999.
8. Dhillon, B.S., Life distributions, *IEEE Transactions on Reliability*, Vol. 30, 1981, pp. 457–459.
9. Shooman, M.L., *Probabilistic Reliability: An Engineering Approach*, McGraw-Hill, New York, 1968.
10. Sandler, G.H., *System Reliability Engineering*, Prentice Hall, Englewood Cliffs, New Jersey, 1963.
11. Lipp, J.P., Topology of switching elements versus reliability, *Transactions on IRE Reliability and Quality Control*, Vol. 7, 1957, pp. 21–34.

4 Reliability Evaluation Methods

4.1 INTRODUCTION

Over the years, a large number of published literature in the area of reliability has appeared in the form of books, journal articles, conference proceeding articles, and technical reports [1–5]. Many of these publications report the development of various types of methods and approaches for performing reliability analysis. Some of these methods are known as the Markov method, fault tree analysis (FTA), and failure modes and effect analysis (FMEA).

The Markov method is named after a Russian mathematician, Andrei A. Markov (1856–1922), and is a highly mathematical approach that is frequently used for evaluating reliability of repairable systems. The FTA method was developed in the early 1960s to analyze the safety of rocket launch control systems in the United States. The FMEA method was developed in the early 1950s by the US Navy's Bureau of Aeronautics [6]. Later, the National Aeronautics and Space Administration (NASA) extended it for classifying each potential failure effect according to its severity and renamed it: failure mode effects and criticality analysis (FMECA) [6]. Nowadays, Markov, FTA, and FMEA methods are being used across many diverse areas to analyze various types of problems.

This chapter presents a number of methods considered useful to evaluate reliability of engineering systems.

4.2 FAILURE MODES AND EFFECT ANALYSIS (FMEA)

FMEA is quite a versatile method widely used in the industrial sector for analyzing systems during the design phase from reliability and safety aspects. It may simply be described as a very effective approach for performing analysis of each and every potential failure mode in the system for determining the effects of such modes on the entire system [7]. The history of this method goes back to the early 1950s when the Bureau of Aeronautics of the US Navy developed a requirement known as failure analysis for establishing a mechanism for reliability control over the detail design-related effort [8]. Subsequently, the term failure analysis was switched over to failure modes and effect analysis.

Usually, the following seven steps are followed for performing FMEA:

- **Step I:** Establish system boundaries and its associated requirements
- **Step II:** List all system, subsystems, and components
- **Step III:** Identify and describe each component/part and list its possible failure modes
- **Step IV:** Assign occurrence probability/failure rate to each failure mode
- **Step V:** List effect(s) of each failure mode on subsystem(s), system, and plant
- **Step VI:** Enter necessary remarks for each identified failure mode
- **Step VII:** Review all critical failure modes and take appropriate actions

It is to be noted that prior to the implementation of FMEA, there are several factors that must be explored carefully. Four of these factors are as follows [9,10]:

- **Factor I:** Examination of each conceivable failure mode by all the involved personnel
- **Factor II:** Making necessary decisions based on the risk priority number
- **Factor III:** Measuring costs/benefits
- **Factor IV:** Obtaining approval and support of the involved engineer(s)

Over the years, professionals involved in reliability analysis have developed a number of guidelines/facts concerning FMEA. Some of these guidelines/facts are as follows [9]:

- Developing the most of FMEA in a meeting should be avoided
- FMEA has certain limitations
- FMEA is not a method for choosing the optimum design concept
- FMEA is not designed for superseding the engineer's work

Some of the main advantages of conducting FMEA are as follows [9,10]:

- Starts from the detailed level and works upward
- Compares designs and highlights safety-related concerns
- Reduces engineering-related changes and improves the efficiency of test planning
- Safeguards against repeating the same mistakes in the future
- A systematic approach to categorize/classify hardware failures
- Helpful to understand and improve customer satisfaction
- Useful visibility tool for management that reduces product development time and cost
- Improves communication between design interface personnel.

4.3   FAULT TREE ANALYSIS (FTA)

This method was developed in the early 1960s at the Bell Telephone Laboratories for performing the safety analysis of the Minuteman Launch Control System [11,12]. Nowadays, it is widely used in industry to evaluate reliability of engineering systems during their design and development phase, particularly in the area of nuclear power generation. A fault tree may simply be described as a logical representation of the relationship of basic fault events that lead to the occurrence of a stated undesirable event, called the "Top event", and is depicted using a tree structure with logic gates such as AND and OR gates.

The main objectives of conducting FTA are as follows [2]:

- To comprehend the functional relationships of system failures
- To satisfy jurisdictional requirements
- To highlight critical areas and cost-effective improvements
- To verify the system's ability to satisfy its imposed safety requirements
- To understand the degree of protection that the design concept provides against failures' occurrence

There are many prerequisites associated with the FTA. Six of the main prerequisites are as follows [2]:

- **I:** Clear identification of all related assumptions
- **II:** Clearly defined what constitutes system/item failure: the undesirable event (top event)
- **III:** Clearly defined analysis scope and objectives
- **IV:** A comprehensive review of system/item operational-related experience
- **V:** Clear comprehension of design, operation, and maintenance aspects of system/item under consideration
- **VI:** Clearly defined system/item interfaces as well as system/item physical bounds.

Four basic symbols used for constructing fault trees are shown in Figure 4.1 [11,12].
Each of the four symbols shown in Figure 4.1 is described below.

- **AND gate:** This symbol denotes that an output fault event occurs only if all the input fault events occur.
- **OR gate:** This symbol denotes that an output fault event occurs if any one or more input fault events occur.
- **Circle:** This symbol denotes a basic or primary fault event (e.g., failure of an elementary component/part) and the basic fault parameters are failure probability, failure rate, repair rate, and unavailability.
- **Rectangle:** This symbol denotes a fault event that results from the logical combination of fault events through the input of a logic gate such as AND and OR.

Normally, FTA is conducted by the following seven steps as shown below [11,13]:

- **Step I:** Define the system and its associated assumptions.

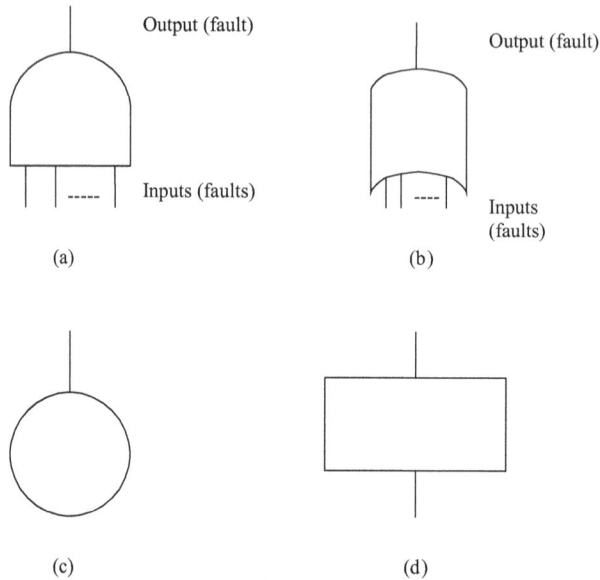

FIGURE 4.1   Basic fault tree symbols: (a) AND gate, (b) OR gate, (c) basic fault event, and (d) resultant fault event.

- **Step II:** Highlight the undesirable fault event (i.e., top fault event) to be investigated.
- **Step III:** Determine all the possible causes that can lead to the occurrence of the top fault event by using fault tree symbols such as those given in Figure 4.1 and the logic tree format.
- **Step IV:** Develop the fault tree to the lowest level of detail as per the requirements.
- **Step V:** Conduct analysis of the developed fault tree in regard to factors such as comprehending the logic and interrelationships among the fault paths and gaining insight into the unique modes of item faults.
- **Step VI:** Determine the most effective corrective measures.
- **Step VII:** Document analysis and follow-up on all highlighted corrective measures.

Example 4.1

Assume that a windowless room has three light bulbs and one switch. Develop a fault tree, using Figure 4.1 symbols, for the undesired (i.e., top) fault event, Dark room, if the switch only fails to close.

In this case, there can only be no light in the room (i.e., Dark room) if all the three light bulbs burn out, if there is no incoming electricity, or if the switch fails to close. Using Figure 4.1 symbols, a fault tree for the example is shown in Figure 4.2. The single capital letters in Figure 4.2 denote corresponding fault events (e.g., T: Dark room [top fault event]).

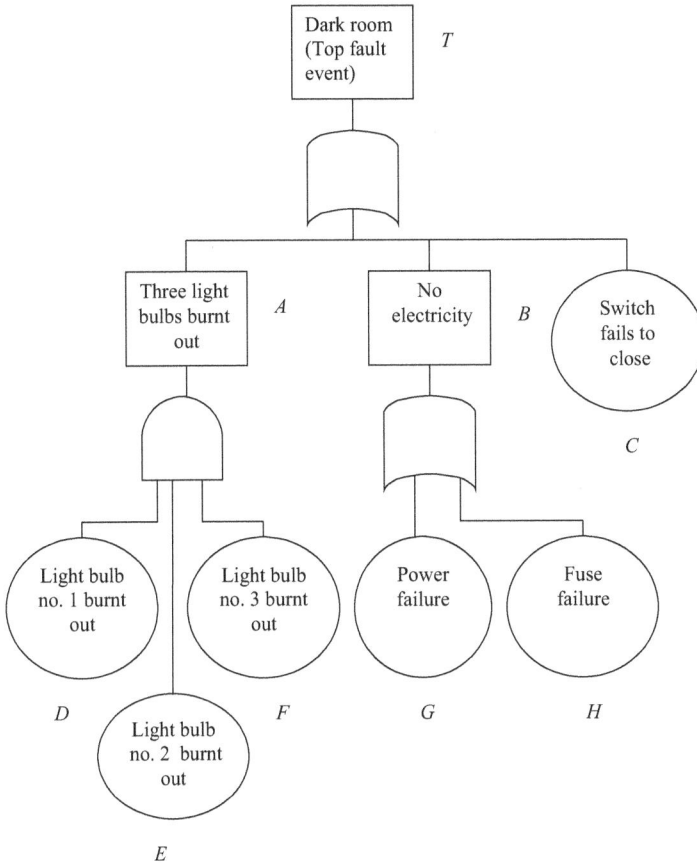

FIGURE 4.2   A fault tree for the top fault event: Dark room.

4.3.1   FAULT TREE PROBABILITY EVALUATION

When occurrence probabilities of basic/primary fault events are known, then the occurrence probability of the top fault event can be calculated. This can only be calculated by first calculating the occurrence probabilities of the output fault events of all the lower and intermediate logic gates (e.g., OR and AND gates).

Thus, the probability of occurrence of the OR gate output fault event, X, is given by [2]

$$P(X) = 1 - \prod_{j=1}^{m} \left[1 - P(X_j) \right] \qquad (4.1)$$

where

$P(X)$ is the occurrence probability of the OR gate output fault event X.
m is the number of OR gate input independent fault events.

$P\left(X_j\right)$ is the occurrence probability of the OR gate input fault event X_j, for $j = 1$, 2, 3,..., m.

Similarly, the probability of occurrence of the AND gate output fault event, Y, is given by [2]

$$P(Y) = \prod_{j=1}^{n} P\left(Y_j\right) \qquad (4.2)$$

where

$P(Y)$ is the occurrence probability of the AND gate output fault event Y.
n is the number of AND gate input independent fault events.
$P\left(Y_j\right)$ is the occurrence probability of the AND gate input fault event Y_j, for $j = 1$, 2, 3,..., n.

Example 4.2

Assume that the occurrence probabilities of independent fault events C, D, E, F, G, and H in Figure 4.2 are 0.01, 0.04, 0.05, 0.06, 0.02, and 0.08, respectively. Calculate the probability of occurrence of the top fault event T (Dark room) using Equations (4.1) and (4.2).

By substituting the given occurrence probability values of fault events G and H into Equation (4.1), we get

$$P(B) = 1 - \left[(1 - 0.02)(1 - 0.08)\right]$$

$$= 0.0984$$

where

$P(B)$ is the occurrence probability of fault event B (No electricity).

Similarly, by substituting the given occurrence probability values of fault events D, E, and F into Equation (4.2), we get

$$P(A) = (0.04)(0.05)(0.06)$$

$$= 0.00012$$

where

$P(A)$ is the occurrence probability value of fault event A (three light bulbs burnt out).

By substituting the above two calculated values and the given data value into Equation (4.1), we get

$$P(T) = 1 - \left[(1-0.00012)(1-0.0984)(1-0.01) \right]$$

$$= 0.8924$$

Thus, the probability of occurrence of the top fault event T (Dark room) is 0.8924. Figure 4.3 shows Figure 4.2 fault tree with the above calculated values and the specified fault event occurrence probability values.

4.4 MARKOV METHOD

This is a widely used method for performing reliability-related analysis of engineering systems and is named after a Russian mathematician, Andrei A. Markov (1856–1922).

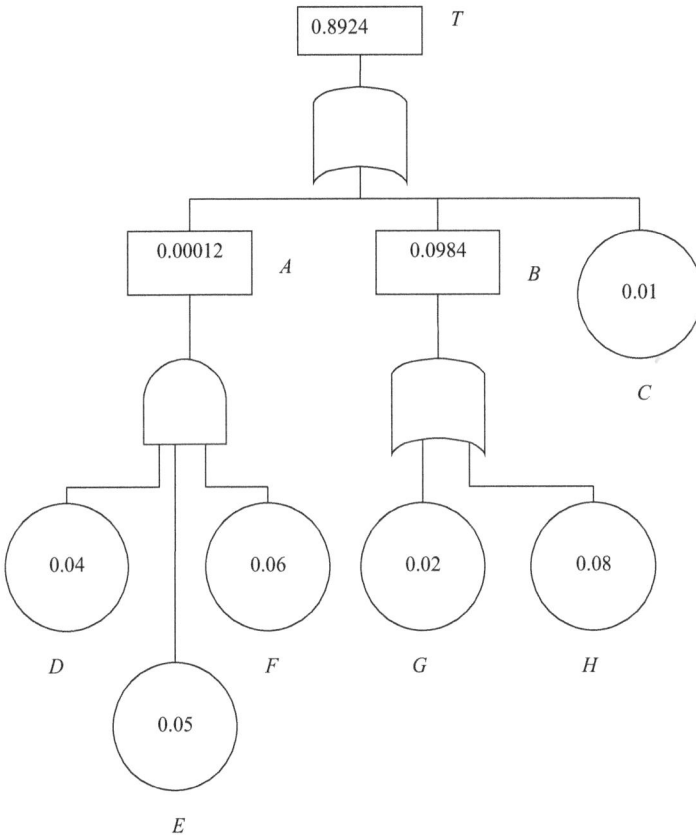

FIGURE 4.3 A fault tree with the given and calculated fault event occurrence probability values.

The method is used quite commonly for modeling repairable engineering systems with constant failure and repair rates and is subject to the following assumptions [14]:

- The transitional probability from one system state to another in the finite time interval Δt is given by $\lambda \Delta t$, where λ is the transition rate (e.g., system failure or repair rate) from one system state to another.
- All occurrences are independent of each other.
- The probability of more than one transition occurrence in the finite time interval Δt from one system state to another is negligible (e.g., $(\lambda \Delta t)(\lambda \Delta t) \to 0$).

The application of this method is demonstrated by solving the example shown below.

Example 4.3

Assume that a system can be either in an operating or a failed state. The system constant failure and repair rates are λ_s and μ_s, respectively. The system state space diagram is shown in Figure 4.4. The numerals in box and diamond denote the system states. Develop equations for the system time-dependent and steady-state availabilities and unavailabilities, reliability, and mean time to failure by using the Markov method.

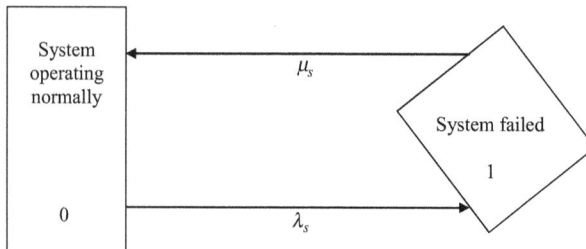

FIGURE 4.4 System state space diagram.

By using the Markov method, we write down the following equations for the system states 0 and 1 shown in Figure 4.4, respectively.

$$P_0(t+\Delta t) = P_0(t)(1 - \lambda_s \Delta t) + P_1(t)\mu_s \Delta t \tag{4.3}$$

$$P_1(t+\Delta t) = P_1(t)(1 - \mu_s \Delta t) + P_0(t)\lambda_s \Delta t \tag{4.4}$$

where

t is the time.
$\lambda_s \Delta t$ is the probability of system failure in finite time interval Δt.
$\mu_s \Delta t$ is the probability of system repair in finite time interval Δt.
$(1 - \lambda_s \Delta t)$ is the probability of no failure in finite time interval Δt.

$\left(1-\mu_s \Delta t\right)$ is the probability of no repair in finite time interval Δt.

$P_j(t)$ is the probability that the system is in state j at time t, for $j = 0, 1$.

$P_0(t+\Delta t)$ is the probability of the system being in operating state 0 at time $(t + \Delta t)$.

$P_1(t+\Delta t)$ is the probability of the system being in failed state 1 at time $(t+\Delta t)$.

From Equation (4.3), we get

$$P_0(t+\Delta t) = P_0(t) - P_0(t)\lambda_s \Delta t + P_1(t)\mu_s \Delta t \qquad (4.5)$$

From Equation (4.5), we write

$$\lim_{\Delta t \to 0} \frac{P_0(t+\Delta t) - P_0(t)}{\Delta t} = -P_0(t)\lambda_s + P_1(t)\mu_s \qquad (4.6)$$

Thus, from Equation (4.6), we obtain

$$\frac{dP_0(t)}{dt} + P_0(t)\lambda_s = P_1(t)\mu_s \qquad (4.7)$$

Similarly, using Equation (4.4), we get

$$\frac{dP_1(t)}{dt} + P_1(t)\mu_s = P_0(t)\lambda_s \qquad (4.8)$$

At time $t = 0$, $P_0(0) = 1$ and $P_1(0) = 0$.

By solving Equations (4.7) and (4.8), we obtain the following equations [2]:

$$P_0(t) = \frac{\mu_s}{\left(\lambda_s + \mu_s\right)} + \frac{\lambda_s}{\left(\lambda_s + \mu_s\right)} e^{-(\lambda_s + \mu_s)t} \qquad (4.9)$$

$$P_1(t) = \frac{\lambda_s}{\left(\lambda_s + \mu_s\right)} - \frac{\lambda_s}{\left(\lambda_s + \mu_s\right)} e^{-(\lambda_s + \mu_s)t} \qquad (4.10)$$

Thus, the system time-dependent availability and unavailability, respectively, are

$$AV_s(t) = P_0(t) = \frac{\lambda_s}{\left(\lambda_s + \mu_s\right)} + \frac{\lambda_s}{\left(\lambda_s + \mu_s\right)} e^{-(\lambda_s + \mu_s)t} \qquad (4.11)$$

and

$$UAV_s(t) = P_1(t) = \frac{\lambda_s}{\left(\lambda_s + \mu_s\right)} - \frac{\lambda_s}{\left(\lambda_s + \mu_s\right)} e^{-(\lambda_s + \mu_s)t} \qquad (4.12)$$

where

$AV_s(t)$ is the system time-dependent availability.
$UAV_s(t)$ is the system time-dependent unavailability.

By letting time t go to infinity in Equations (4.11) and (4.12), we get [2]

$$AV_s = \lim_{t \to \infty} AV_s(t) = \frac{\mu_s}{\lambda_s + \mu_s} \qquad (4.13)$$

and

$$UAV_s = \lim_{t \to \infty} UAV_s(t) = \frac{\lambda_s}{\lambda_s + \mu_s} \qquad (4.14)$$

where

AV_s is the system steady-state availability.
UAV_s is the system steady-state unavailability.

For $\mu_s = 0$, from Equation (4.9), we get

$$R_s(t) = P_0(t) = e^{-\lambda_s t} \qquad (4.15)$$

By integrating Equation (4.15) over the time interval $[0, \infty]$, we get the following equation for the system mean time to failure [2]:

$$MTTF_s = \int_0^\infty e^{-\lambda_s t}\, dt$$

$$= \frac{1}{\lambda_s} \qquad (4.16)$$

where

$MTTF_s$ is the system mean time to failure.

Thus, the system time-dependent and steady-state availabilities and unavailabilities, reliability, and mean time to failure are given by Equations (4.11), (4.13), (4.12), (4.14), (4.15), and (4.16), respectively.

Example 4.4

Assume that the constant failure and repair rates of a system used in industry are 0.0004 failures per hour and 0.0008 repairs per hour, respectively. Calculate the system steady-state availability and availability during a 100-hour mission.

By substituting the given data into Equations (4.13) and (4.11), we get

$$AV_s = \frac{0.0008}{0.0004 + 0.0008} = 0.6666$$

and

$$AV_s(100) = \frac{0.0008}{(0.0004 + 0.0008)} + \frac{(0.0004)}{(0.0004 + 0.0008)} e^{-(0.0004 + 0.0008)(100)}$$

$$= 0.9623$$

Thus, the system steady-state availability and availability during a 100-hour mission are 0.6666 and 0.9623, respectively.

4.5 NETWORK REDUCTION APPROACH

This is probably the simplest approach for determining the reliability of systems composed of independent series and parallel systems. However, the subsystems forming bridge networks/configurations can also be handled by first using the delta-star method [15]. Nonetheless, the network reduction approach sequentially reduces the series and parallel subsystems to equivalent hypothetical single units until the whole system under consideration itself becomes a single hypothetical unit. The example presented below demonstrates this approach.

Example 4.5

An independent unit network representing a system is shown in Figure 4.5 (i). The reliability R_j of unit j; for j = 1, 2, 3, ..., 7 is given. Calculate the network reliability by utilizing the network reduction approach.

First, we have highlighted subsystems A, B, C, and D of the network as shown in Figure 4.5 (i). The subsystems B and C have their units in series; thus, we reduce them to single hypothetical units as follows:

$$R_B = R_4 R_5 = (0.5)(0.9) = 0.45$$

and

$$R_C = R_6 R_7 = (0.4)(0.3) = 0.12$$

where

R_B is the subsystem B reliability.
R_C is the subsystem C reliability.

The reduced network is shown in Figure 4.5 (ii). Now, the network is made up of two parallel subsystems A and D. Thus, we reduce both these subsystems to single hypothetical units as follows:

$$R_A = 1-(1-R_1)(1-R_2)(1-R_3)$$
$$= 1-(1-0.8)(1-0.6)(1-0.7)$$
$$= 0.976$$

and

$$R_D = 1-(1-R_B)(1-R_C)$$
$$= -1-(1-0.45)(1-0.12)$$
$$= 0.516$$

where

R_A is the subsystem A reliability.
R_D is the subsystem D reliability.

Figure 4.5 (iii) shows the reduced network with the above calculated values. This resulting network is a two-unit series system and its reliability is given by

$$R_S = R_A R_D = (0.976)(0.516)$$
$$= 0.5036$$

The single hypothetical unit shown in Figure 4.5 (iv) represents the reliability of the whole network shown in Figure 4.5 (i). More clearly, the whole network is reduced to a single hypothetical unit. Thus, the whole network reliability, R_S, is 0.5036.

4.6 DECOMPOSITION APPROACH

This approach is used for determining reliability of complex systems, which it decomposes into simpler subsystems by using the conditional probability theory. Subsequently, the system reliability is determined by combining reliability measures of the subsystems.

The basis for this approach is the selection of the key unit used for decomposing a given network. The approach's efficiency depends on the selection of this key unit. The past experience normally plays a pivotal role in its selection.

The approach/method starts with the assumption that the key unit, say *m*, is replaced by another unit that never fails (i.e., 100% reliable) and then it assumes

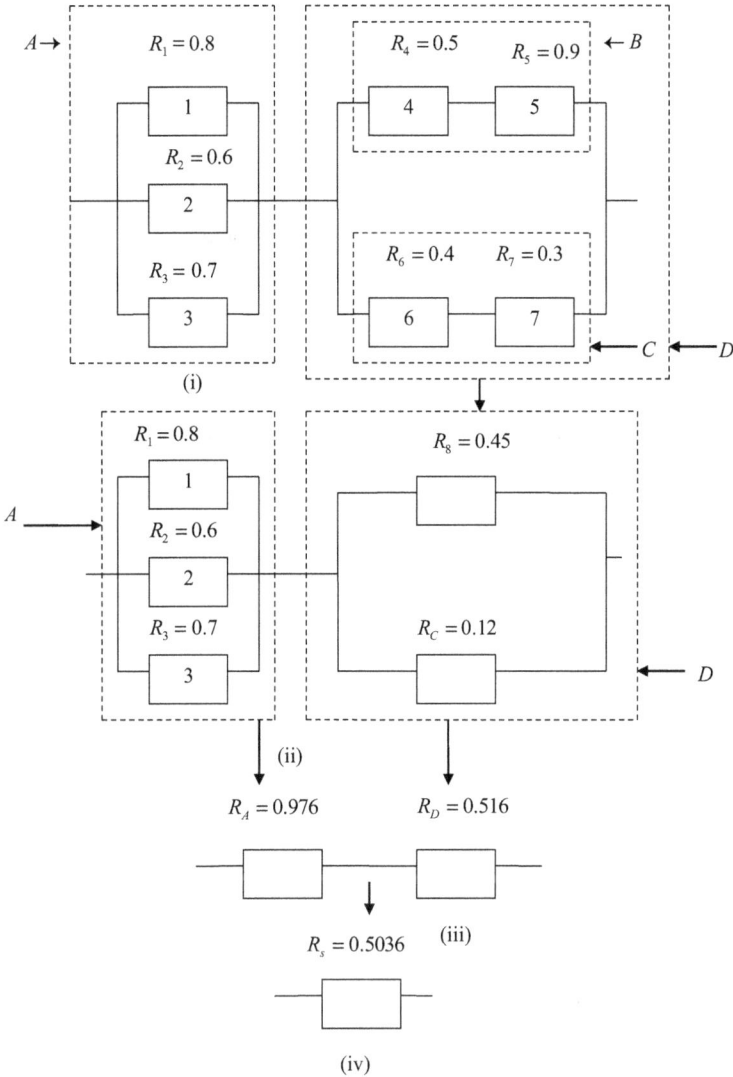

FIGURE 4.5 Diagrammatic steps of the network reduction approach: (i) original network; (ii) reduced network; (iii) reduced network; (iv) single hypothetical unit.

that the key unit is completely removed from the system/network. Thus, the overall system/network reliability is expressed by

$$R_s = P(m)P(\text{system good} / m \text{ good}) + P(\bar{m}) \, P(\text{system good} / m \text{ fails}) \qquad (4.17)$$

where

R_S is the overall system/network reliability.
$P(m)$ is the probability of success or reliability of the key unit m.
$P(\bar{m})$ is the probability of failure or unreliability of the key unit m.
$P(.)$ is the probability.

The application of this approach/method is demonstrated by solving the following example.

Example 4.6

Assume that five independent and identical units form a bridge network/system as shown in Figure 4.6. The capital letter R in the figure denotes unit reliability. Obtain an expression for reliability of the bridge network/system by using the decomposition approach/method.

In this case with the aid of past experience, we choose the unit falling between nodes X and Y, shown in Figure 4.6, as our key unit, say m.

Next, we replace this key unit m with a unit that is 100% reliable (i.e., never fails). Consequently, the network shown in Figure 4.6 becomes a series–parallel network/system whose reliability is given by

$$R_{spn} = \left[1-(1-R)^2\right]^2$$
$$= \left[2R - R^2\right]^2 \tag{4.18}$$

where

R_{spn} is the series–parallel network reliability.

Similarly, we totally remove the key unit m from Figure 4.6 and the resulting network becomes a parallel-series network. This parallel-series network reliability is expressed by

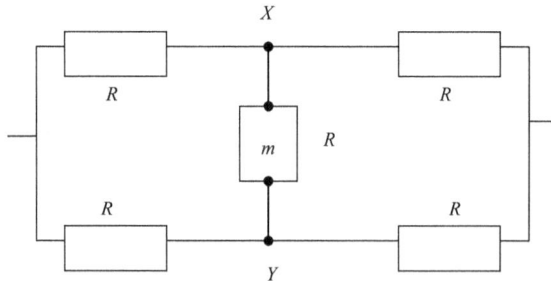

FIGURE 4.6　A five identical units bridge network/system.

$$R_{psn} = 1 - \left(1 - R^2\right)^2$$

$$= 2R^2 - R^4 \qquad (4.19)$$

where

R_{psn} is the parallel-series network reliability.

The reliability and unreliability of the key unit m, respectively, are expressed by

$$P(m) = R \qquad (4.20)$$

and

$$P(\bar{m}) = \left(1 - R\right) \qquad (4.21)$$

Rewriting Equation (4.17) in terms of this example (i.e., Example 4.6), we obtain

$$R_s = RR_{spn} + \left(1 - R\right)R_{psn} \qquad (4.22)$$

By inserting Equations (4.18) and (4.19) into Equation (4.22), we get

$$R_s = R\left(2R - R^2\right)^2 + \left(1 - R\right)\left(2R^2 - R^4\right)$$

$$= 2R^5 - 5R^4 + 2R^3 + 2R^2 \qquad (4.23)$$

Equation (4.23) is for the reliability of the bridge network/system shown in Figure 4.6.

4.7   DELTA-STAR METHOD

This is the simplest and a practical method for evaluating reliability of independent units bridge networks. The method transforms a bridge network to its equivalent parallel and series form. However, it is to be noted that the transformation process introduces a minor error in the end result, but for practical purposes it should be neglected [16].

Once a bridge network is transformed to its equivalent series and parallel form, the network reduction approach can be used for obtaining network reliability. Nonetheless, the delta-star method can quite easily handle networks containing more than one bridge configurations. Furthermore, it can also be applied to bridge configurations composed of devices having two mutually exclusive failure modes [5,15].

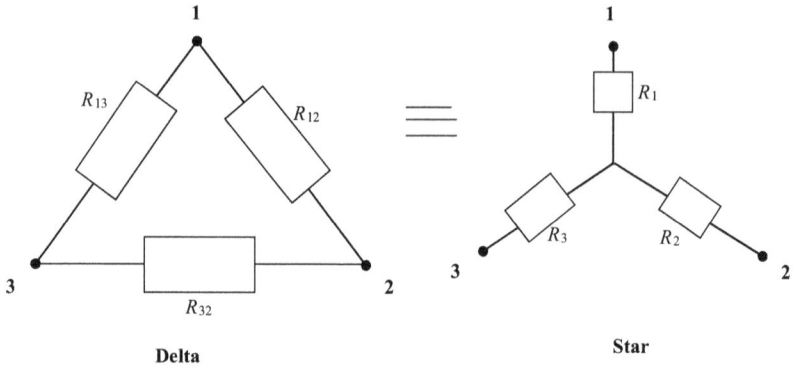

FIGURE 4.7 Delta to star equivalent reliability diagram.

Delta to star equivalent reliability diagram is shown in Figure 4.7. In this diagram, the numbers 1, 2, and 3 denote nodes, the blocks the units, and $R(.)$ the respective unit reliability.

The delta-star equivalent legs are shown in Figure 4.8.

Reliabilities of Figure 4.8 delta to star equivalent diagrams (i), (ii), and (iii), respectively, are as follows:

$$R_1 R_2 = 1 - \left(1 - R_{12}\right)\left(1 - R_{13} R_{32}\right) \tag{4.24}$$

$$R_2 R_3 = 1 - \left(1 - R_{32}\right)\left(1 - R_{12} R_{13}\right) \tag{4.25}$$

$$R_1 R_3 = 1 - \left(1 - R_{13}\right)\left(1 - R_{32} R_{12}\right) \tag{4.26}$$

By solving Equations (4.24)–(4.26), we obtain

$$R_1 = \sqrt{XZ/Y} \tag{4.27}$$

where

$$X = 1 - \left(1 - R_{12}\right)\left(1 - R_{13} R_{32}\right) \tag{4.28}$$

$$Y = 1 - \left(1 - R_{32}\right)\left(1 - R_{12} R_{13}\right) \tag{4.29}$$

$$Z = 1 - \left(1 - R_{13}\right)\left(1 - R_{32} R_{12}\right) \tag{4.30}$$

$$R_2 = \sqrt{XY/Z} \tag{4.31}$$

$$R_3 = \sqrt{YZ/X} \tag{4.32}$$

Delta   **Star**

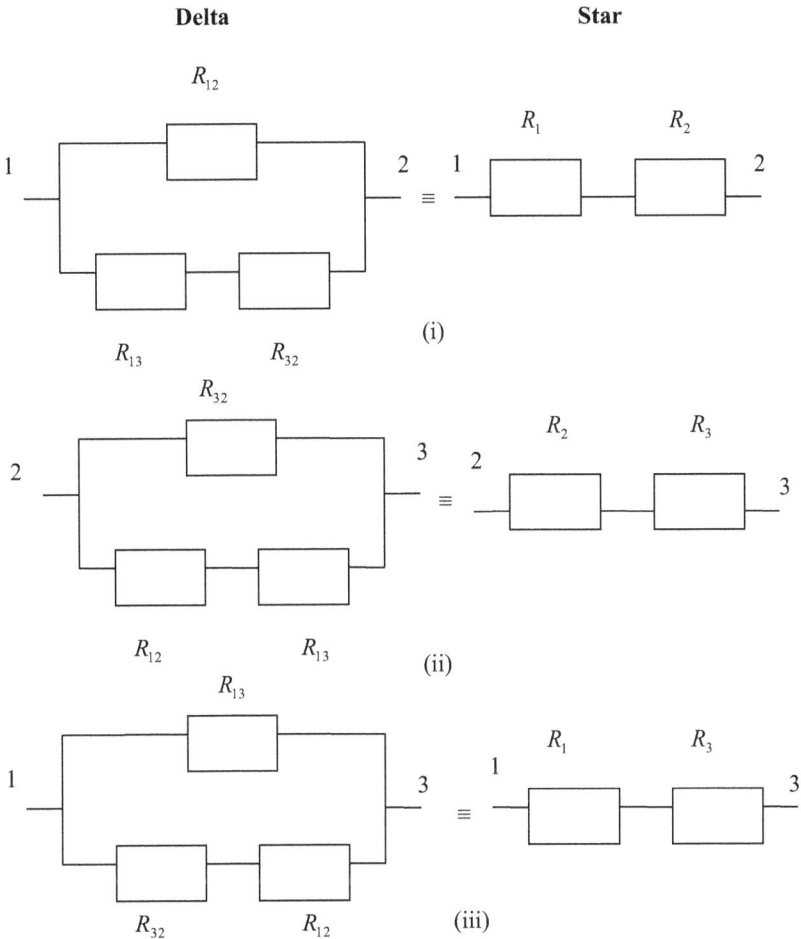

(i)

(ii)

(iii)

FIGURE 4.8 Delta to star equivalent diagrams (i), (ii), (iii).

Example 4.7

A five independent unit bridge network with given unit reliability R_j, for j = a, b, c, d, and e is shown in Figure 4.9. Calculate the network reliability by using the delta-star method and also use the stated data values in Equation (4.23) for obtaining the bridge network reliability. Compare both results.

In Figure 4.9, nodes 1, 2, and 3 denote delta configuration. By using Equations (4.27)–(4.32) and the stated data values, we obtain the following star equivalent units' reliabilities:

$$R_1 = \sqrt{XZ/Y}$$
$$= 0.9904$$

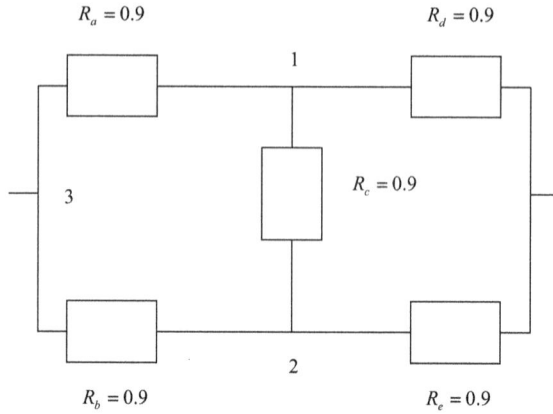

FIGURE 4.9   A five unit bridge network with given unit reliabilities.

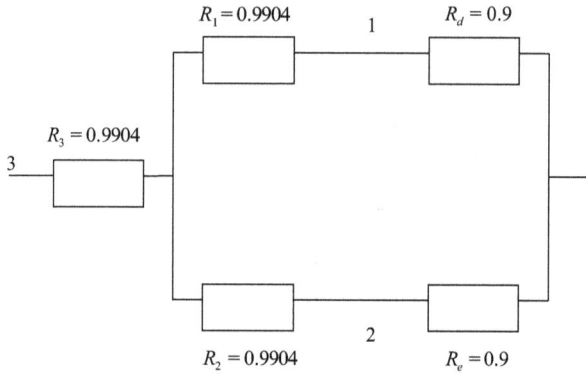

FIGURE 4.10   Equivalent network to bridge network shown in Figure 4.9.

where

$$X = Y = Z = 1 - (1 - 0.9)[1 - (0.9)(0.9)]$$

$$= 0.981$$

$$R_2 = 0.9904$$

and

$$R_3 = 0.9904$$

Using the above results, the equivalent network to Figure 4.9 bridge network is shown in Figure 4.10.

The reliability of network in Figure 4.10 is, R_{bn},

$$R_{bn} = R_3 \left[1 - \left(1 - R_1 R_d \right) \left(1 - R_2 R_e \right) \right]$$
$$= 0.9787$$

By inserting the specified data values into Equation (4.23), we obtain

$$R_s = 2(0.9)^5 - 5(0.9)^4 + 2(0.9)^3 + 2(0.9)^2$$
$$= 0.9784$$

It is to be noted that both the above reliability results are basically same, i.e., 0.9787 and 0.9784. It basically means that for practical purposes the delta-star method is quite effective.

4.8 PROBABILITY TREE ANALYSIS

This method can be used for performing reliability-related task analysis by diagrammatically representing human actions and other associated events in question. In this case, diagrammatic task analysis is represented by the branches of the probability tree. More specifically, the tree's branching limbs represent each event's outcome (i.e., success or failure) and each branch is assigned probability of occurrence [17].

Some of the advantages of this method are flexibility for incorporating (i.e., with some modifications) factors such as interaction effects, emotional stress, and interaction stress; simplified mathematical computations; and a visibility tool. It is to be noted that the method can also be used for evaluating reliability of networks such as series, parallel, and series–parallel. The method's application to such configurations is demonstrated in Dhillon [18].

Nonetheless, the following example demonstrates the application of this method:

Example 4.8

Assume that a person has to perform two independent and distinct tasks m and n to operate an engineering system. Task m is performed before task n. Furthermore, each of these two tasks can be conducted either correctly or incorrectly.

Develop a probability tree and obtain an equation for probability of not successfully accomplishing the overall mission (i.e., not operating the engineering system correctly) by the person.

In this example, the person first performs task m correctly or incorrectly and then proceeds to perform task n. This task can also be performed either correctly or incorrectly. Figure 4.11 depicts a probability tree for the entire scenario.

The symbols used in the figure are defined below.

m denotes the event that task m is performed correctly.
$\bar{m}$ denotes the event that task m is performed incorrectly.

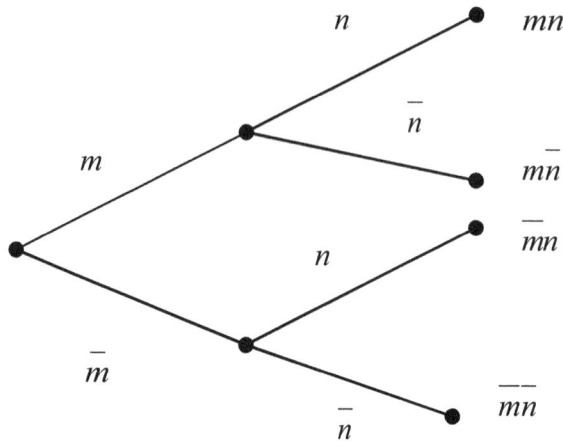

FIGURE 4.11 A probability tree for performing tasks m and n.

n denotes the event that task n is performed correctly.
$\bar{n}$ denotes the event that task n is performed incorrectly.

In Figure 4.11, the term mn denotes operating the engineering system successfully (i.e., overall mission success). Thus, the probability of occurrence of events mn is given by [18]

$$P(mn) = P_m P_n \tag{4.33}$$

where

P_m is the probability of performing task m correctly.
P_n is the probability of performing task n correctly.

Similarly, in Figure 4.11, the terms $m\bar{n}$, $\bar{m}n$, and $\bar{m}\bar{n}$ denote three distinct possibilities of not operating the engineering system correctly or successfully. Thus, the probability of not successfully accomplishing the overall mission (i.e., not operating the engineering system correctly) by the person is

$$P_f = P(m\bar{n} + \bar{m}n + \bar{m}\bar{n})$$
$$= P_m P_{\bar{n}} + P_{\bar{m}} P_n + P_{\bar{m}} P_{\bar{n}} \tag{4.34}$$

where

$P_{\bar{n}}$ is the probability of performing task n incorrectly.
$P_{\bar{m}}$ is the probability of performing task m incorrectly.

P_f is the probability of not successfully accomplishing the overall mission (i.e., mission failure).

Example 4.9

Assume that in Example 4.8, the probabilities of the person not performing tasks m and n correctly are 0.3 and 0.25, respectively. Calculate the probability of correctly operating the engineering system by the person.

Thus, we have $P_{\bar{m}} = 0.3$ and $P_{\bar{n}} = 0.25$.

Since $P_{\bar{m}} + P_m = 1$ and $P_{\bar{n}} + P_n = 1$, we have

$$P_m = 1 - P_{\bar{m}} \qquad (4.35)$$

$$P_n = 1 - P_{\bar{n}} \qquad (4.36)$$

By substituting Equations (4.35)–(4.36) and the specified data values into Equation (4.33), we obtain

$$P(mn) = (1 - P_{\bar{m}})(1 - P_{\bar{n}})$$

$$= (1 - 0.3)(1 - 0.25)$$

$$= 0.525$$

Thus, probability of correctly operating the engineering system (i.e., the probability of occurrence of events mn) by the person is 0.525.

4.9 BINOMIAL METHOD

This method is used for evaluating the reliability of relatively simple systems, such as series and parallel systems/networks. For such systems'/networks' reliability evaluation, this is one of the simplest methods. However, in the case of complex systems/networks the method becomes a trying task. The method can be applied to systems/networks with independent identical or non-identical units. The following formula is the basis for the method [16]:

$$\prod_{i=1}^{n}(R_i + F_i) \qquad (4.37)$$

where

n is the number of non-identical units/components.
R_i is the ith unit reliability.
F_i is the ith unit failure probability.

Example 4.10

Using Equation (4.37) develop reliability expressions for parallel and series networks having two non-identical and independent units each.

In this case, since $n = 2$ from Equation (4.37) we get

$$\left(R_1 + F_1\right)\left(R_2 + F_2\right) = R_1 R_2 + R_1 F_2 + R_2 F_1 + F_1 F_2 \qquad (4.38)$$

where

R_1 is the reliability of unit 1.
R_2 is the reliability of unit 2.
F_1 is the failure probability of unit 1.
F_2 is the failure probability of unit 2.

Thus, using Equation (4.38), we write the following reliability expression for the parallel network having two non-identical units:

$$R_{p2} = R_1 R_2 + R_1 F_2 + R_2 F_1 \qquad (4.39)$$

where

R_{p2} is the two non-identical units parallel network reliability.

Since $\left(R_1 + F_1\right) = 1$ and $\left(R_2 + F_2\right) = 1$, Equation (4.39) becomes

$$R_{p2} = R_1 R_2 + R_1 \left(1 - R_2\right) + R_2 \left(1 - R_1\right) \qquad (4.40)$$

By rearranging Equation (4.40) we obtain

$$R_{p2} = R_1 R_2 + R_1 - R_1 R_2 + R_2 - R_1 R_2$$

$$= R_1 + R_2 - R_1 R_2$$

$$= 1 - \left(1 - R_1\right)\left(1 - R_2\right) \qquad (4.41)$$

Finally, the two non-identical units series network reliability from Equation (4.38) is

$$R_{s2} = R_1 R_2 \qquad (4.42)$$

where

R_{s2} is the two non-identical units series network reliability.

Thus, reliability expressions for parallel and series networks having two non-identical and independent units each are given by Equations (4.41) and (4.42), respectively.

4.10 PROBLEMS

1. Describe failure modes and effect analysis method and its advantages.
2. What are the main objectives of conducting FTA and the main prerequisites associated with FTA?
3. Assume that a windowless room has five light bulbs and one switch. Develop a fault tree for the undesired (i.e., top) fault event, *Dark room*, if the switch only fails to close.
4. What are the assumptions associated with the Markov method?
5. Assume that the constant failure and repair rates of an engineering system are 0.0005 failures per hour and 0.0007 repairs per hour, respectively. Calculate the engineering system steady-state unavailability and unavailability during a 50-hour mission.
6. Assume that five independent and identical units form a bridge network and reliability of each unit is 0.8. Calculate the network reliability by using the delta-star method and the decomposition approach. Compare both the results.
7. Assume that a person has to perform three independent and distinct tasks a, b, and c to operate an engineering system. Task a is performed before task b, and task b before task c. Furthermore, each of these tasks can be performed either correctly or incorrectly. Develop a probability tree and obtain an expression for probability of not successfully accomplishing the overall mission (i.e., not operating the engineering system correctly) by the person.
8. Compare probability tree analysis with FTA.
9. Describe binomial method and write down its basic formula.
10. Develop reliability expression for a parallel network with three independent and non-identical units by using the binomial method.

REFERENCES

1. Grant Ireson, W., Coombs, C.F., Moss, R.Y., Editors, *Handbook of Reliability Engineering and Management*, McGraw-Hill, New York, 1996.
2. Dhillon, B.S., *Design Reliability: Fundamentals and Applications*, CRC Press, Boca Raton, Florida, 1999.
3. RDG-376, *Reliability Design Handbook, Reliability Analysis Center*, Rome Air Development Center, Griffis Air Force Base, Rome, New York, 1976.
4. AMCP 706-196, *Engineering Design Handbook: Development Guide for Reliability, Part II: Design for Reliability*, US Army Material Command (AMC), Washington, DC, 1976.
5. Dhillon, B.S., Proctor, C.L., Reliability Analysis of Multistate Device Networks, *Proceedings of the Annual Reliability and Maintainability Symposium*, 1976, pp. 31–35.
6. Jordan, W.E., Failure Modes, Effects, and Criticality Analyses, *Proceedings of the Annual Reliability and Maintainability Symposium*, 1972, pp. 30–37.

7. Omdahl, T.P., Editor, *Reliability, Availability, and Maintainability (RAM) Dictionary*, American Society for Quality Control (ASQC) Press, Milwaukee, Wisconsin, 1988.

8. MIL-F-18372 (Aer), *General Specification for Design, Installation, and Test of Aircraft Flight Control Systems*, Bureau of Naval Weapons, Department of the Navy, Washington, DC.

9. Palady, P., *Failure Modes and Effects Analysis*, PT Publications, West Palm Beach, Florida, 1995.

10. McDermott, R.E., Mikulak, K.J., Beauregard, M.R., *The Basics of FMEA*, Quality Resources, New York, 1996.

11. Dhillon, B.S., Singh, C., *Engineering Reliability: New Techniques and Applications*, John Wiley, New York, 1981.

12. Schroder, R.J., Fault Tree for Reliability Analysis, *Proceedings of the Annual Symposium on Reliability*, 1970, pp. 170–174.

13. Mears, P., *Quality Improvement Tools and Techniques*, McGraw Hill, New York, 1995.

14. Shooman, M.L., *Probabilistic Reliability: An Engineering Approach*, McGraw Hill, New York, 1968.

15. Dhillon, B.S., *The Analysis of the Reliability of Multi-State Device Networks*, Ph.D. Dissertation, 1975. Available from the National Library of Canada, Ottawa.

16. Dhillon, B.S., *Reliability in Systems Design and Operation*, Van Nostrand Reinhold, New York, 1983.

17. Swain, A.D., *A Method for Performing a Human Factors Reliability Analysis*, Report No. SCR-685, Sandia Corporation, Albuquerque, New Mexico, USA, August 1963.

18. Dhillon, B.S., *Human Reliability: With Human Factors*, Pergamon Press, New York, 1986.

5 Robot Reliability

5.1 INTRODUCTION

Robots are increasingly being used for performing various types of tasks including arc welding, spot welding, routing, and materials handling. A robot may simply be described as a mechanism guided by automatic controls and the word "robot" is derived from the Czechoslovakian language, in which it means "worker" [1].

The first commercial robot was manufactured by the Planet Corporation in 1959 [2]. Nowadays, millions of industrial robots are being used around the globe [3]. As robots use electrical, mechanical, hydraulic, pneumatic, and electronic components, their reliability-associated problems are highly challenging because of many different sources of failures. Although there is no clear-cut definitive point in the beginning of robot reliability field, a publication by JF Engelberger, in 1974, could be regarded as its starting point [4]. A comprehensive list of publications on robot reliability is available in Refs. [5,6].

This chapter presents various important aspects of robot reliability.

5.2 ROBOT FAILURE CLASSIFICATIONS, CAUSES, AND CORRECTIVE MEASURES

Robot failures can be categorized under the four classifications as shown in Figure 5.1 [7–9].

Classification I (i.e., human errors) failures occur due to the personnel who design, manufacture, test, operate, and maintain robots. Some of the causes for the occurrence of human errors are as follows:

- Poor equipment/system design
- Task complexities
- Poorly written operating and maintenance procedures
- Poor training of operating and maintenance personnel
- High temperature in the work area
- Improper tools
- Inadequate lighting in the work area

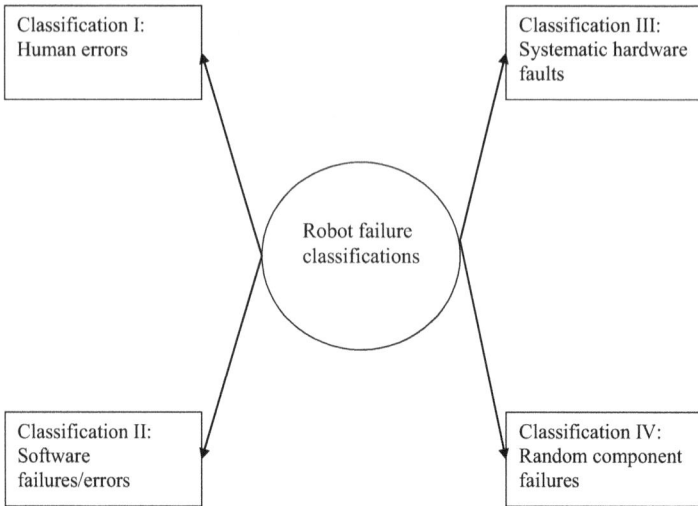

FIGURE 5.1   Robot failure classifications.

Thus, human errors may be divided into categories such as inspection errors, design errors, installation errors, operating errors, assembly errors, and maintenance errors. Some of the methods that can be used to reduce the occurrence of human errors are as follows:

- Fault tree analysis
- Man–machine systems analysis
- Quality circles
- Error cause removal program

The first method (i.e., fault tree analysis) is described in Chapter 4 and the remaining three methods are described in Dhillon [10].

Classification II (i.e., software failures/errors) failures are associated with robot software and in robots, software faults/failures/errors can happen in the embedded software or the controlling software and application software. Redundancy, even though it is expensive, is probably the best solution to protect against the occurrence of software failures/errors. Also, the application of approaches such as fault tree analysis, failure modes and effect analysis, and testing can be quite useful for reducing the occurrence of software failures/errors. Furthermore, there are many software reliability models that can also be used for evaluating reliability when the software in question is put into operational use [11–13].

Classification III (i.e., systematic hardware faults) failures are those failures that can take place due to unrevealed mechanisms present in the robot system design. Some of the reasons for the occurrence of such faults are unusual joint-to-straight-line mode transition, failure to make the appropriate environment-associated provisions in the initial design, and peculiar wrist orientations.

Some of the methods that can be employed for reducing the occurrence of robot-related systematic hardware failures are the use of sensors for detecting the loss of pneumatic pressure, line voltage, or hydraulic pressure; and the employment of sensors for detecting excessiveness of temperature, acceleration, speed, force, and servo errors. Several methods considered quite useful for reducing systematic hardware failures are described in Dhillon [11].

Finally, Classification IV (i.e., random component failures) failures are those failures that occur unpredictably during the useful life of components. Some of the reasons for the occurrence of such failures are undetectable defects, unavoidable failures, low safety factors, and unexplainable causes. The methods presented in Chapter 4 can be used for reducing the occurrence of such failures.

5.3 ROBOT RELIABILITY–RELATED SURVEY RESULTS AND ROBOT EFFECTIVENESS DICTATING FACTORS

Jones and Dawson [14] reported the results of a robot reliability study that was based on surveys of 37 robots of 4 different design used in 3 different companies X, Y, and Z; covering 21,932 robot production hours. These three companies (i.e., X, Y, and Z) reported 47, 306, and 155 cases of robot reliability–related problems, respectively, of which the corresponding 27, 35, and 1 cases did not contribute to any downtime. More specifically, robot downtime as a proportion of production time for these three companies (i.e., X, Y, and Z) was 1.8%, 13.6%, and 5.1, respectively.

Approximate mean time to robot failure (MTTRF) and mean time to robot-related problems (MTTRP) in hours for companies X, Y, and Z are shown in Figure 5.2.

It is to be noted that as shown in Figure 5.2, among these three companies, there is a wide variation of MTTRF and MTTRP. More specifically, the highest to lowest *MTTRF* and *MTTRP* levels are 2596.40 and 221.15 hours, respectively.

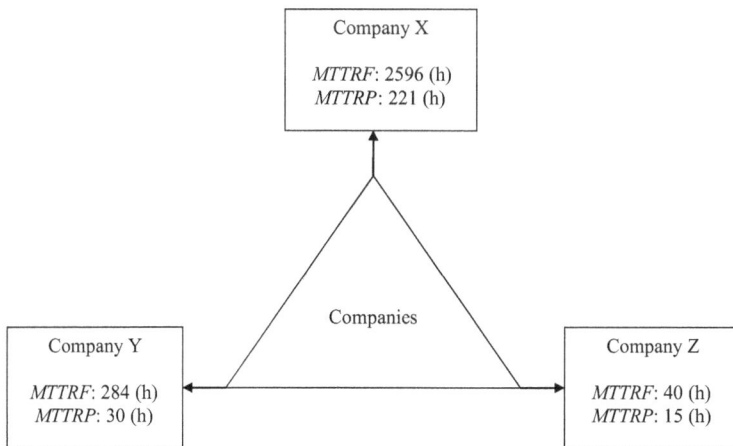

FIGURE 5.2 Approximate mean time to robot failure (MTTRF) and mean time to robot-related problems (MTTRP) in hours (h) for companies *X*, *Y*, and *Z*.

There are many factors that dictate the effectiveness of robots. Some of these factors are as follows [9,15]:

- Percentage of time the robot operates normally
- Mean time between failures of the robot
- Mean time to repair of the robot
- Availability and quality of manpower needed for keeping the robot in operating state
- Percentage of time the robot operates normally
- Relative performance of the robot under extreme conditions
- Availability and quality of the robot repair equipment and facilities
- Rate of the availability of the needed spare parts/components

5.4 ROBOT RELIABILITY MEASURES

There are various types of reliability measures. Four of these measures are as follows [9,11,16]:

5.4.1 ROBOT RELIABILITY

Robot reliability may simply be defined as the probability that a robot will perform its stated function satisfactorily for the specified time period when used as per designed conditions. The general formula for obtaining time-dependent robot reliability is expressed by [9,11].

$$R_r(t) = \exp\left[-\int_0^t \lambda_r(t)\,dt\right]$$ (5.1)

where

$R_r(t)$ is the robot reliability at time t.
$\lambda_r(t)$ is the time-dependent failure rate (hazard rate) of the robot.

Equation (5.1) can be used to obtain the reliability function of a robot for any failure times probability distribution (e.g., exponential, Rayleigh, or Weibull).

Example 5.1

Assume that the time-dependent failure rate of a robot is defined by

$$\lambda_r(t) = \frac{2t}{\alpha^2}$$ (5.2)

where

α is the distribution parameter.
t is the time.

$\lambda_r(t)$ is the hazard rate (time-dependent failure rate) of the robot when its times to failure follow the Rayleigh distribution.

Obtain an expression for the robot reliability.
    By inserting Equation (5.2) into Equation (5.1), we obtain

$$R_r(t) = \exp\left[-\int_0^t \frac{2t}{\alpha^2}\,dt\right]$$

$$= e^{-\left(\frac{t}{\alpha}\right)^2} \tag{5.3}$$

Thus, Equation (5.3) is the expression for the robot reliability.

Example 5.2

Assume that the constant failure rate of a robot is 0.0005 failures per hour. Calculate the robot reliability for a 10 hours mission.
    By inserting the robot specified failure rate value into Equation (5.1), we get

$$R_r(t) = \exp\left[-\int_0^t (0.0005)\,dt\right]$$

$$= e^{-(0.0005)t} \tag{5.4}$$

Substituting the stated mission time value of the robot into Equation (5.4) yields

$$R_r(10) = e^{-(0.0005)(10)}$$

$$= 0.9950$$

Thus, the robot reliability for the stated mission period is 0.9950.

5.4.2 MEAN TIME TO ROBOT FAILURE (MTTRF)

MTTRF can be obtained by using any of the following three equations:

$$MTTRF = \int_0^\infty R_r(t)\,dt \tag{5.5}$$

$$MTTRF = \lim_{s \to 0} R_r(s) \tag{5.6}$$

$$MTTRF = \frac{RPH - DDTRF}{NRF} \tag{5.7}$$

where

MTTRF is the mean time to robot failure.
$R_r(t)$ is the robot reliability at time t.
$R_r(s)$ is the Laplace transform of the robot reliability function, $R_r(t)$.
s is the Laplace transform variable.
RPH is the robot production hours.
NRF is the number of robot failures.
DDTRF is the downtime due to robot failures expressed in hours.

Example 5.3

Assume that the constant failure rate of a robot is 0.0008 failures per hour and its reliability is expressed by

$$R_r(t) = e^{-\lambda_r t}$$

$$= e^{-(0.0008)t} \tag{5.8}$$

where

$R_r(t)$ is the robot reliability at time t.
λ_r is the robot constant failure rate.

Calculate the mean time to robot failure by using Equations (5.5) and (5.6) and comment on the final result.
By substituting Equation (5.8) into Equation (5.5), we obtain

$$MTTRF = \int_0^\infty e^{-(0.0008)t} \, dt$$

$$= \frac{1}{0.0008}$$

$$= 1250 \text{ hours}$$

By taking the Laplace transform of Equation (5.8), we get

$$R_r(s) = \frac{1}{(s + 0.0008)} \tag{5.9}$$

By inserting Equation (5.9) into Equation (5.6), we obtain

$$MTTRF = \lim_{s \to 0} \frac{1}{(s + 0.0008)}$$

$$= \frac{1}{(0.0008)}$$

$$= 1250 \text{ hours}$$

In both cases, the end result (i.e., $MTTRF = 1250$ hours) is the same. It proves that both equations (i.e., Equations (5.5) and (5.6)) yield exactly the same end result.

Example 5.4

Assume that a robot's annual production hours and its annual downtime due to failures are 5000 hours and 250 hours, respectively. During that period, the robot failed four times. Calculate the mean time to robot failure.

By substituting the given data values into Equation (5.7), we get

$$MTTRF = \frac{5000 - 250}{4}$$

$$= 1187.5 \text{ hours}$$

Thus, the mean time to robot failure is 1187.5 hours.

5.4.3 ROBOT HAZARD RATE

The robot hazard rate or time-dependent failure rate is expressed by [9,11]

$$\lambda_r(t) = -\frac{1}{R_r(t)} \cdot \frac{dR_r(t)}{dt} \tag{5.10}$$

where

$\lambda_r(t)$ is the robot hazard rate (i.e., time-dependent failure rate).
$R_r(t)$ is the robot reliability at time t.

It is to be noted that Equation (5.10) can be used to obtain the hazard rate of a robot when its times to failure follow any time-continuous probability distribution (e.g., exponential, Weibull, and Rayleigh).

Example 5.5

Assume that the reliability of a robot is expressed by

$$R_r(t) = e^{-\left(\frac{t}{\theta}\right)^2} \qquad (5.11)$$

where

$R_r(t)$ is the robot reliability at time t.
θ is the distribution parameter.

Obtain an expression for hazard rate of the robot.
   By inserting Equation (5.11) into Equation (5.10), we obtain

$$\lambda_r(t) = -\frac{1}{e^{-\left(\frac{t}{\theta}\right)^2}} \cdot \frac{de^{-\left(\frac{t}{\theta}\right)^2}}{dt}$$

$$= -\frac{1}{e^{-\left(\frac{t}{\theta}\right)^2}} \left[-e^{-\left(\frac{t}{\theta}\right)^2} \frac{2t}{\theta^2} \right]$$

$$= \frac{2t}{\theta^2} \qquad (5.12)$$

Thus, Equation (5.12) is the expression for the hazard rate of the robot.

5.4.4   MEAN TIME TO ROBOT PROBLEMS

This is the average productive robot time before the occurrence of a robot-related problem. It is expressed by

$$MTTRP = \frac{RPH - DDTRP}{NRP} \qquad (5.13)$$

where

NRP is the number of robot-related problems.
RPH is the robot production hours.
$DDTRP$ is the downtime due to robot-related problems expressed in hours.
$MTTRP$ is the mean time to robot-related problems.

Example 5.6

Assume that at an industrial facility the annual robot production hours and downtime due to robot-associated problems are 4000 hours and 200 hours, respectively. During the one-year period, there were 20 robot-associated problems. Calculate the mean time to robot problems.

By inserting the specified data values into Equation (5.13), we obtain

$$MTTRP = \frac{4000 - 200}{20}$$

$$= 190 \text{ hours}$$

Thus, the mean time to robot problems is 190 hours.

5.5 RELIABILITY ANALYSIS OF ELECTRIC AND HYDRAULIC ROBOTS

As both electric and hydraulic robots are used in the industry, this section presents reliability analysis of two typical electric and hydraulic robots by employing the block diagram approach/method [7–9]. Generally, for the purpose of design evaluation in the industrial sector, it is assumed for both electric and hydraulic robots that all robot parts form a series configuration (i.e., if any one part/component fails, the robot fails).

5.5.1 RELIABILITY ANALYSIS OF THE ELECTRIC ROBOT

An electric robot considered here is the one that conducts a "normal" industrial task, while its maintenance and programming are conducted by humans. The robot is subject to the following seven assumptions/factors [8,9]:

i   Supervising controller/computer directs all joints.
ii   Transducer sends all appropriate signals to the joint controller.
iii Microprocessor control card controls each and every joint.
iv   Interface bus permits interaction between the supervisory controller and the joint control processors.
v   Motor shaft rotation is transmitted to the appropriate limb of the robot through a transmission unit.
vi   Each and every joint is coupled with a feedback encoder (transducer).
vii Direct current (DC) motor actuates each joint.

In regard to reliability, the block diagram shown in Figure 5.3 represents the electric robot under consideration.

Figure 5.3 shows that the electric robot under consideration is made up of two hypothetical subsystems 1 and 2 in series. Subsystem 1 represents no movement due

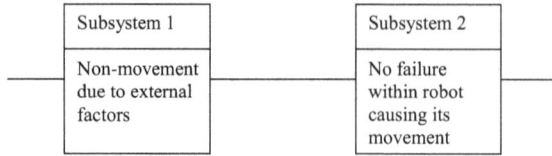

FIGURE 5.3 Block diagram for estimating the non-occurrence probability (reliability) of the undesirable movement of the electric robot.

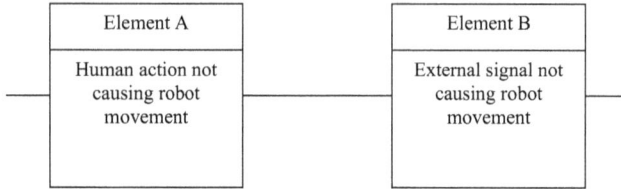

(a)

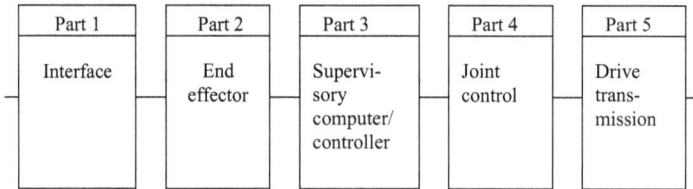

(b)

FIGURE 5.4 Block diagram representing two subsystems shown in Figure 5.3: (a) subsystem 1, (b) subsystem 2.

to external factors, and subsystem 2 represents no failure within the robot causing its movement.

In turn, as shown in Figure 5.4(a), Figure 5.3 subsystem 1 is composed of two hypothetical elements A and B in series and subsystem 2 is made of five parts (i.e., interface, end effector, supervisory controller/computer, joint control, and transmission) (Figure 5.4(b)) in series.

Furthermore, the Figure 5.4(a) element A is composed of two hypothetical subelements X and Y in series as shown in Figure 5.5.

With the aid of Figure 5.3, we obtain the following equation for the probability of non-occurrence of the undesirable electric robot movement (i.e., reliability):

$$R_{em} = R_{s1}R_{s2} \tag{5.14}$$

Subelement X	Subelement Y
Maintenance person's reliability in regard to causing robot movement	Operator's reliability in regard to causing robot movement

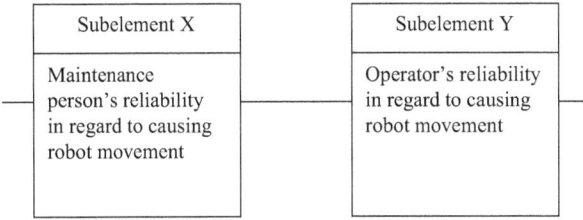

FIGURE 5.5   Block diagram representing Figure 5.4 (a) element A.

where

R_{em} is the probability of non-occurrence (reliability) of the undesirable electric robot movement.

R_{s1} is the independent subsystem 1's reliability.

R_{s2} is the independent subsystem 2's reliability.

For independent elements A and B, the reliability of subsystem 1 in Figure 5.4(a) is given by

$$R_{s1} = R_A R_B \qquad (5.15)$$

where

R_A is the element A's reliability.

R_B is the element B's reliability.

For hypothetical and independent subelements, the element A's reliability in Figure 5.5 is

$$R_A = R_X R_Y \qquad (5.16)$$

where

R_X is the reliability of subelement X (i.e., the maintenance person's reliability in regard to causing the robot's movement).

R_Y is the reliability of subelement Y (i.e., the operator's reliability in regard to causing the robot's movement).

Similarly, the reliability of subsystem 2 in Figure 5.4(b), for independent parts, is given by

$$R_{s2} = R_{jc} R_{ef} R_{dt} R_{sc} R_i \qquad (5.17)$$

where

R_{jc} is the reliability of the joint control.
R_{ef} is the reliability of the end-effector.
R_{dt} is the reliability of the drive transmission.
R_{sc} is the reliability of the supervisory controller/computer.
R_i is the reliability of the interface.

Example 5.7

Assume that the following reliability data values are specified for the above type of electric robot:

$R_B = 0.95$
$R_Y = 0.93$
$R_X = 0.90$
$R_{jc} = 0.94$
$R_{ef} = 0.92$
$R_{dt} = 0.96$
$R_{sc} = 0.91$
$R_i = 0.97$

Calculate the probability of non-occurrence (reliability) of the undesirable electric robot movement.

By inserting the given data values into Equation (5.16) and (5.17), we obtain

$$R_A = (0.90)(0.93)$$
$$= 0.837$$

and

$$R_{s2} = (0.94)(0.0.92)(0.96)(0.91)(0.97)$$
$$= 0.7328$$

By substituting the above-calculated value for R_A and the specified value for R_B into Equation (5.15), we obtain

$$R_{s1} = (0.837)(0.95)$$
$$= 0.7951$$

By inserting the above-calculated values into Equation (5.14), we get

$$R_{em} = (0.7951)(0.7328)$$
$$= 0.5826$$

Thus, the probability of non-occurrence (reliability) of the undesirable electric robot movement is 0.5826.

5.5.2 RELIABILITY ANALYSIS OF THE HYDRAULIC ROBOT

A hydraulic robot considered here is composed of five joints and, in turn, each joint is controlled and driven by a hydraulic servomechanism. The robot is subject to the following seven assumptions/factors [7,9]:

i   Hydraulic fluid is pumped from the reservoir.
ii Under high flow demand, an accumulator assists the pump for supplying additional hydraulic fluid.
iii Unloading valve is employed for keeping pressure under the maximum limit.
iv Position transducer provides the joint angle codes and, in turn, the scanning of each code is conducted by a multiplexer.
v   Servo valve controls the motion of each hydraulic actuator. This motion is transmitted directly or indirectly (i.e., through gears, chains, rods, etc.) to the robot's specific limb and, in turn, each limb is coupled to a position transducer.
vi Operator makes use of a teach pendant for controlling the arm-motion in teach mode.
vii Conventional motor and pump assembly generates pressure.

The hydraulic robot under consideration with respect to reliability is represented by the block diagram shown in Figure 5.6. This figure shows that the hydraulic robot is composed of four subsystems: subsystem 1 (electronic and control subsystem), subsystem 2 (hydraulic pressure supply subsystem), subsystem 3 (gripper subsystem), and subsystem 4 (drive subsystem), in series. In turn, as shown in Figure 5.7 hydraulic pressure supply subsystem (i.e., block diagram (a)) is composed of two parts (i.e., piping and hydraulic equipment/component) in series and gripper subsystem (i.e., block diagram (b)) is also composed of two parts (i.e., control signal and pneumatic system) in series.

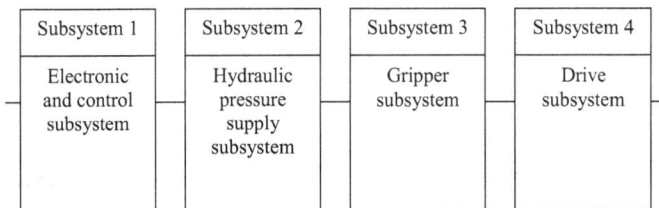

Subsystem 1	Subsystem 2	Subsystem 3	Subsystem 4
Electronic and control subsystem	Hydraulic pressure supply subsystem	Gripper subsystem	Drive subsystem

FIGURE 5.6 Block diagram of the hydraulic robot under consideration.

(a)

(b)

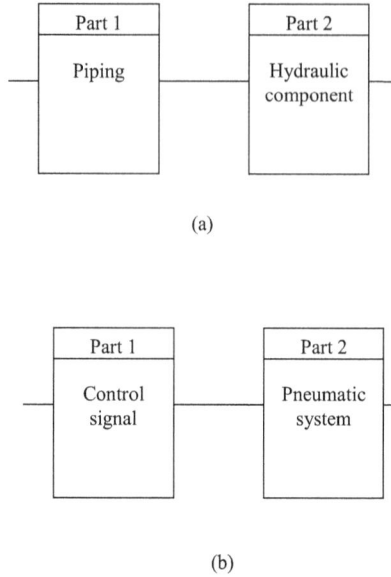

FIGURE 5.7 Block diagram representing two subsystems shown in Figure 5.6: (a) hydraulic pressure supply subsystem and (b) gripper subsystem.

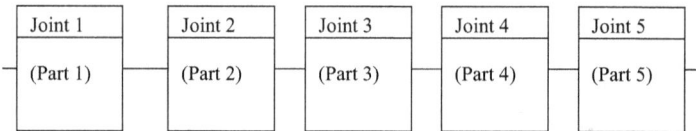

FIGURE 5.8 Block diagram representing subsystem 4 (i.e., drive subsystem) shown in Figure 5.6.

Furthermore, as shown in the Figure 5.8, the drive subsystem (shown in Figure 5.6) is composed of five parts (i.e., joints 1, 2, 3, 4, and 5) in series.

With the aid of Figure 5.6, we get the following expression for the probability of the non-occurrence of the hydraulic robot event (i.e., undesirable hydraulic robot movement causing damage to the robot other equipment as well as possible harm to humans):

$$R_{hr} = R_e R_h R_g R_d \tag{5.18}$$

where

R_e is the reliability of the independent electronic and control subsystem.
R_h is the reliability of the independent hydraulic pressure supply subsystem.
R_g is the reliability of the independent gripper subsystem.

R_d is the reliability of the independent drive subsystem.

R_{hr} is the hydraulic robot reliability or the probability of the non-occurrence of the hydraulic robot event (i.e., undesirable robotic arm movement causing damage to the robot/other equipment as well as possible harm to humans).

For independent parts, the reliabilities $R_h, R_g,$ and R_d of hydraulic pressure supply subsystem, gripper subsystem, and drive subsystem, using Figures 5.7(a), 5.7(b), and 5.8, respectively, are

$$R_h = R_p R_{hc}$$ (5.19)

$$R_g = R_{cs} R_{ps}$$ (5.20)

and

$$R_d = \prod_{j=1}^{5} R_j$$ (5.21)

where

R_p is the reliability of the piping.
R_{hc} is the reliability of the hydraulic component.
R_{cs} is the reliability of the control signal.
R_{ps} is the reliability of the pneumatic system.
R_j is the reliability of joint j; for $j = 1, 2, 3, 4, 5$.

For constant failure rates of independent subsystems shown in Figure 5.6, in turn, of their independent parts shown in Figure 5.7 and Figure 5.8; from Equations (5.18) through (5.21), we get

$$R_{hr}(t) = e^{-\lambda_e t} e^{-\lambda_h t} e^{-\lambda_g t} e^{-\lambda_d t}$$

$$R_{hr}(t) = e^{-\lambda_e t} e^{\lambda_p t} e^{-\lambda_{hc} t} e^{-\lambda_{cs} t} e^{-\lambda_{ps} t} e^{-\sum_{i=1}^{5} \lambda_i t}$$

$$= e^{-\left(\lambda_e + \lambda_p + \lambda_{hc} + \lambda_{cs} + \lambda_{ps} + \sum_{i=1}^{5} \lambda_i\right) t}$$ (5.22)

where

λ_e is the constant failure rate of the electronic and control subsystem.
λ_h is the constant failure rate of the hydraulic pressure supply subsystem.
λ_g is the constant failure rate of the gripper subsystem.

λ_d is the constant failure rate of the drive subsystem.
λ_p is the constant failure rate of the piping.
λ_{hc} is the constant failure rate of the hydraulic component.
λ_{cs} is the constant failure rate of the control signal.
λ_{ps} is the constant failure rate of the pneumatic system.
λ_i is the constant failure rate of the joint i, for $i = 1, 2, 3, 4, 5$.

By integrating Equation (5.22) over the time interval $[0, \infty]$, we obtain

$$MTTOHRUE = \int_0^\infty e^{-\left(\lambda_e + \lambda_p + \lambda_{hc} + \lambda_{cs} + \lambda_{ps} + \sum_{i=1}^{5} \lambda_i\right)t} dt$$

$$= \frac{1}{\left(\lambda_e + \lambda_p + \lambda_{hc} + \lambda_{cs} + \lambda_{ps} + \sum_{i=1}^{5} \lambda_i\right)} \tag{5.23}$$

where

MTTOHRUE is the mean time to the occurrence of the hydraulic robot undesirable event (i.e., undesirable arm movement causing damage to the robot/other equipment and possible harm to humans).

Example 5.8

Assume that the constant failure rates of the above type of hydraulic robot are $\lambda_e = 0.0008$ failures/hour, $\lambda_p = 0.0007$ failures/hour, $\lambda_{hc} = 0.0006$ failures/hour, $\lambda_{cs} = 0.0005$ failures/hour, $\lambda_{ps} = 0.0004$ failures/hour, and $\lambda_1 = \lambda_2 = \lambda_3 = \lambda_4 = \lambda_5 = 0.0003$ failures/hour. Calculate the mean time to the occurrence of the hydraulic robot undesirable event (i.e., undesirable arm movement causing damage to the robot/other equipment and possible harm to humans).

By inserting the given data values into Equation (5.23), we obtain

$$MTTOHRUE = \frac{1}{0.0008 + 0.0007 + 0.0006 + 0.0005 + 0.0004 + 5(0.0003)}$$
$$= 222.2 \text{ hours}$$

Thus, the mean time to the occurrence of the hydraulic robot undesirable event (i.e., undesirable arm movement causing damage to the robot/other equipment and possible harm to humans) is 222.2 hours.

5.6 MODELS FOR PERFORMING ROBOT RELIABILITY AND MAINTENANCE STUDIES

There are many mathematical models that can be used, directly or indirectly, to conduct various types of robot reliability and maintenance studies. Three of these models are presented below.

5.6.1 MODEL I

This model is concerned with determining the economic life of a robot. More clearly, the time limit beyond this is not economical to conduct robot repairs. Thus, the robot economic life is defined by [16–19]:

$$REL = \left[\frac{2(RIC - RSV)}{RRC_{ai}} \right]^{1/2} \tag{5.24}$$

where

REL is the robot economic life.
RIC is the robot initial cost (installed).
RSV is the robot scrap value.
RRC_{ai} is the robot's annual increase in repair cost.

Example 5.9

Assume that an electrical robot costs $150,000 (installed) and its estimated scrap value is $3,000. The estimated annual increase in its repair cost is $200. Estimate the time limit beyond which the robot-related repairs will not be beneficial.

By substituting the given data values into Equation (5.24), we obtain

$$REL = \left[\frac{2(150,000 - 3,000)}{200} \right]^{1/2}$$

$$= 38.34 \text{ years}$$

Thus, the time limit beyond which the robot-related repairs will not be beneficial is 38.34 years.

5.6.2 MODEL II

This model represents a robot system that can fail either due to a human error or other failures (e.g., hardware and software) and the failed robot system is repaired to its operating state. The robot system state–space diagram is shown in Figure 5.9.

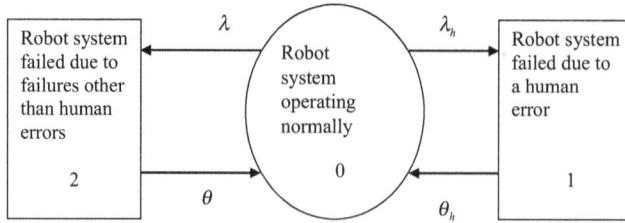

FIGURE 5.9   Robot system state–space diagram.

The numerals in the diagram circle and rectangles denote system states. The model is subjected to the following assumptions [9,19]:

- Human error and other failure rates are constant.
- Failed robot system repair rates are constant.
- Human error and other failures are statistically independent and the repaired robot system is as good as new.

The following symbols are associated with the diagram shown in Figure 5.9 and its associated equations:

$P_j(t)$ is the probability that the robot system is in state j at time t; for $j = 0$ (operating normally), $j = 1$ (failed due to a human error), $j = 2$ (failed due to failures other than human errors).
λ is the robot system constant non-human error failure rate.
λ_h is the robot system constant human error rate.
θ is the robot system constant repair rate from failed state 2.
θ_h is the robot system constant repair rate from failed state 1.

Using the Markov method presented in Chapter 4, we write down the following equations for Figure 5.9 [9,19]:

$$\frac{dP_0(t)}{dt} + (\lambda + \lambda_h) P_0(t) = \theta P_2(t) + \theta_h P_1(t) \tag{5.25}$$

$$\frac{dP_1(t)}{dt} + \theta_h P_1(t) = \lambda_h P_0(t) \tag{5.26}$$

$$\frac{dP_2(t)}{dt} + \theta P_2(t) = \lambda P_0(t) \tag{5.27}$$

At time $t = 0$, $P_0(0) = 1, P_1(0) = 0, and, P_2(0) = 0.$

Solving Equations (5.25)–(5.27) using Laplace transforms, we obtain

$$P_0(t) = \frac{\theta\theta_h}{n_1 n_2} + \left[\frac{(n_1 + \theta)(n_1 + \theta_h)}{n_1(n_1 - n_2)} \right] e^{n_1 t} \left[\frac{(n_2 + \theta)(n_2 + \theta_h)}{n_2(n_1 - n_2)} \right] e^{n_2 t} \qquad (5.28)$$

where

$$n_1, n_2 = \frac{-a \pm \left[a^2 - 4(\theta\theta_h + \lambda_h\theta + \lambda\theta_h) \right]^{1/2}}{2} \qquad (5.29)$$

$$a = \lambda + \lambda_h + \theta + \theta_h \qquad (5.30)$$

$$n_1 n_2 = \theta\theta_h + \lambda_h\theta + \lambda\theta_h \qquad (5.31)$$

$$(n_1 + n_2) = (\lambda + \lambda_h + \theta + \theta_h) \qquad (5.32)$$

$$P_1(t) = \frac{\theta\lambda_h}{n_1 n_2} + \left[\frac{\lambda_h n_1 + \lambda_h\theta}{n_1(n_1 - n_2)} \right] e^{n_1 t} - \left[\frac{(\theta + n_2)\lambda_h}{n_2(n_1 - n_2)} \right] e^{n_2 t} \qquad (5.33)$$

$$P_2(t) = \frac{\lambda\theta_h}{n_1 n_2} + \left[\frac{\lambda n_1 + \lambda\theta_h}{n_1(n_1 - n_2)} \right] e^{n_1 t} - \left[\frac{(\theta_h + n_2)\lambda}{n_2(n_1 - n_2)} \right] e^{n_2 t} \qquad (5.34)$$

The robot system availability, $AV_{rs}(t)$, is given by

$$AV_{rs}(t) = P_0(t) \qquad (5.35)$$

As time t becomes large in Equations (5.33)–(5.35), we obtain the following steady-state probability equations:

$$AV_{rs} = \frac{\theta\theta_h}{n_1 n_2} \qquad (5.36)$$

$$P_1 = \frac{\theta\lambda_h}{n_1 n_2} \qquad (5.37)$$

$$P_2 = \frac{\lambda\theta_h}{n_1 n_2} \qquad (5.38)$$

where

AV_{rs} is the robot system steady-state availability.
P_1 is the steady-state probability of the robot system being in state 1.
P_2 is the steady-state probability of the robot system being in state 2.

For $\theta = \theta_h = 0$, from Equations (5.28), (5.33), and (5.34) we obtain

$$P_0(t) = e^{-(\lambda + \lambda_h)t} \tag{5.39}$$

$$P_1(t) = \frac{\lambda_h}{(\lambda + \lambda_h)}\left[1 - e^{-(\lambda + \lambda_h)t}\right] \tag{5.40}$$

$$P_2(t) = \frac{\lambda}{(\lambda + \lambda_h)}\left[1 - e^{-(\lambda + \lambda_h)t}\right] \tag{5.41}$$

The robot system reliability from Equation (5.39) is

$$R_{rs}(t) = e^{-(\lambda + \lambda_h)t} \tag{5.42}$$

where

$R_{rs}(t)$ is the robot system reliability at time t.

By inserting Equation (5.42) into Equation (5.5), we obtain the following equation for *MTTRF*:

$$MTTRF = \int_0^\infty e^{-(\lambda + \lambda_h)}dt$$

$$= \frac{1}{(\lambda + \lambda_h)} \tag{5.43}$$

Using Equation (5.42) in Equation (5.10), we obtain the following equation for the robot system hazard rate:

$$\lambda_r = -\frac{1}{e^{-(\lambda + \lambda_h)t}} \cdot \frac{d\left[e^{-(\lambda + \lambda_h)t}\right]}{dt}$$

$$= \lambda + \lambda_h \tag{5.44}$$

It is to be noted that the right-hand side of Equation (5.44) is independent of time t, which means the failure rate of the robot system is constant.

Example 5.10

Assume that a robot system can fail either due to a human error or other failures, and its human errors and other failure rates are 0.0002 errors per hour and 0.0006 failures per hour, respectively. The robot system repair rate from both the failure modes is 0.004 repairs per hour. Calculate the robot system steady-state availability.

By inserting the given data values into Equation (5.36), we get

$$AV_{rs} = \frac{(0.004)(0.004)}{(0.004)(0.004)+(0.0002)(0.004)+(0.0006)(0.004)} = 0.8333$$

Thus, the robot system steady-state availability is 0.8333.

5.6.3 MODEL III

This mathematical model can be used for calculating the optimum number of inspections per robot facility per unit time [18,19]. This information is very useful to decision makers but inspections are quite often disruptive; however, such inspections generally reduce the robot downtime because they lower breakdowns. In this model, the total downtime of the robot is minimized to get the optimum number of inspections.

The total downtime of the robot, *TDTR*, per unit time is expressed by [20]

$$TDTR = mT_{di} + \frac{cT_{db}}{m} \tag{5.45}$$

where

    m is the number of inspections per robot facility per unit time.
    c is a constant for a specific robot facility.
    T_{di} is the downtime per inspection for a robot facility.
    T_{db} is the downtime per breakdown for a robot facility.

By differentiating Equation (5.45) with respect to m and then equating it to zero, we obtain

$$m^* = \left[\frac{cT_{db}}{T_{di}} \right]^{1/2} \tag{5.46}$$

where

    m^* is the optimum number of inspections per robot facility per unit time.

By inserting Equation (5.46) into Equation (5.45), we obtain

$$TDTR^* = 2\left[T_{di}.cT_{db}\right]^{1/2} \tag{5.47}$$

where

$TDTR^*$ is the minimum total downtime of the robot.

Example 5.11

Assume that for a robot facility, the following data values are given:

- $c = 2$
- $T_{db} = 0.1$ months
- $T_{di} = 0.02$ months

Calculate the optimum number of robot inspections per month and the minimum total robot downtime.

By inserting the above specified values into Equations (5.46) and (5.47), we obtain

$$m^* = \left[\frac{2(0.1)}{0.02}\right]^{1/2}$$
$$= 3.16 \text{ inspections per month}$$

and

$$TDTR^* = 2\left[(0.02)(2)(0.1)\right]^{1/2}$$
$$= 0.12 \text{ months}$$

Thus, the optimum number of robot inspections per month and the minimum total robot downtime are 3.16 and 0.12 months, respectively.

5.7 PROBLEMS

1. Discuss classifications of robot failures and their causes.
2. What are the robot effectiveness dictating factors?
3. Write down the general formula for obtaining time-dependent robot reliability.
4. Write down formula for calculating mean time to robot problems.
5. Compare an electric robot with a hydraulic robot with respect to reliability.
6. Assume that an electric robot costs $200,000 (installed) and its estimated scrap value is $4,000. The estimated annual increase in its repair cost is $250.

Estimate the time limit beyond which the robot-related repairs will not be beneficial.

7. Assume that a robot system can fail either due to a human error or other failures, and its human error and other failure rates are 0.0004 errors per hour and 0.0008 failures per hour, respectively. The repair rate of the robot system from both the failure modes is 0.005 repairs per hour. Calculate the robot system steady-state unavailability.
8. Prove Equation (5.47) by using Equation (5.45).
9. Prove that the sum of Equations (5.28), (5.33), and (5.34) is equal to unity.
10. Assume that for a robot facility, the following data values are given:

- T_{db} = 0.2 months
- T_{di} = 0.01 months
- $c = 3$

Calculate the optimum number of robot inspections per month and the minimum total robot downtime.

REFERENCES

1. Jablonowski, J., Posey, J.W., *Robotics Terminology*, in Handbook of Industrial Robotics, edited by S.Y. Nof, John Wiley, New York, 1985, pp. 1271–1303.
2. Zeldman, M.I., *What Every Engineer Should Know About Robots*, Marcel Dekker, New York, 1984.
3. Rudall, B.H., Automation and robotics worldwide: reports and surveys, *Robotica*, Vol. 14, 1996, pp. 164–168.
4. Engleberger, J.F., Three million hours of robot field experience, *The Industrial Robot*, 1974, pp. 164–168.
5. Dhillon, B.S., On robot reliability and safety: bibliography, *Microelectronics and Reliability*, Vol. 27, 1987, pp. 105–118.
6. Dhillon, B.S., Fashandi, A.R.M., Liu, K.L., Robot systems reliability and safety: a review, *Journal of Quality in Maintenance Engineering*, Vol. 18, No. 3, 2002, pp. 170–212.
7. Khodanbandehloo, K., Duggan, F., Husband, T.F., Reliability Assessment of Industrial Robots, *Proceedings of the 14th International Symposium on Industrial Robots*, 1984, pp. 209–220.
8. Khodanbandehloo, K., Duggan, F., Husband, T.F., Reliability of Industrial Robots: A Safety Viewpoint, Proceedings of Industrial Robots: A Safety Viewpoint, *Proceedings of the 7th British Robot Association Annual Conference*, 1984, pp. 233–242.
9. Dhillon, B.S., *Robot Reliability and Safety*, Springer-Verlag, New York, 1991.
10. Dhillon, B.S., *Human Reliability: With Human Factors*, Pergamon Press, New York, 1986.
11. Dhillon, B.S., *Design Reliability: Fundamentals and Applications*, CRC Press, Boca Raton, FL, 1999.
12. Herrmann, D.S., *Software Safety and Reliability*, IEEE Computer Society Press, Los Alamitos, CA, 1999.
13. Dhillon, B.S., *Reliability in Computer Systems Design*, Ablex Publishing, Norwood, NJ, 1987.
14. Jones, R., Dawson, S., People and Robots: Their Safety and Reliability, *Proceedings of the 7th British Robot Association Annual Conference*, 1984, pp. 243–258.

15. Young, J.F., *Robotics*, Butterworth, London, 1973.
16. Varnum, E.C., Bassett, B.B., Machine and Tool Replacement Practices, in Manufacturing Planning and Estimating Handbook, edited by F. W. Wilson and P. D. Harvey, McGraw Hill, New York, 1963, pp. 18.1–18.22.
17. Eidmann, F.L., *Economic Control of Engineering and Manufacturing*, McGraw Hill, New York, 1931.
18. Dhillon, B.S., *Mechanical Reliability: Theory, Models, and Applications*, American Institute of Aeronautics and Astronautics, Washington, DC, 1988.
19. Dhillon, B.S., *Applied Reliability and Quality*, Springer-Verlag, London, 2007.
20. Wild, R., *Essential of Production and Operations Management*, Holt, Reinhart, and Winston, London, 1985, pp. 356–368.

6 Computer and Internet Reliability

6.1 INTRODUCTION

Nowadays, billions of dollars are being spent annually around the globe to produce computers for various types of applications ranging from personal use to control space and other systems. The computers are composed of both hardware and software components and for their successful operation, the reliability of both these components is equally important. The history of computer hardware reliability may be traced back to the late 1940s and 1950s [1–4]. For example, the triple modular scheme for improving computer hardware reliability was proposed by Von Neumann in 1956 [4]. It appears that the first serious effort on software reliability started in 1964 at Bell Laboratories [5]. However, some of the important works that appeared in the 1960s on software reliability are provided in Refs. [5–7].

The history of the Internet may be traced back to 1969 with the development of Advanced Research Project Agency Network (ARPANET) [8]. It has grown from 4 hosts in 1969 to about 147 million hosts and 38 million sites in 2002, and nowadays billions of people around the globe use Internet services [8]. In 2001, there were over 52,000 Internet-associated failures and incidents. Needless to say, the reliability and stability of the Internet has become very important to the global economy and other areas, because Internet-related failures can cause millions of dollars in losses and interrupt the day-to-day routines of millions of end users around the globe [9]. This chapter presents various important aspects of computer hardware, software, and Internet reliability.

6.2 COMPUTER FAILURE CAUSES AND ISSUES IN COMPUTER SYSTEM RELIABILITY

There are many causes of computer failures. The important ones are as follows [10–12]:

- Processor and memory failures/errors
- Communication network failures
- Peripheral device failures
- Environmental and power failures

- Human errors
- Mysterious failures
- Saturation
- Gradual erosion of the database

The first six of the above causes of computer failures are described below [10–12]:

- **Processor and memory failures/errors**: These failures/errors are generally catastrophic, but their occurrence is quite rare, as there are times when the central processor malfunctions and fails to execute instructions correctly due to a "dropped bit". Nowadays, the occurrence of memory parity errors is very rare because of improvements in hardware reliability, and these errors are not necessarily fatal.
- **Communication network failures**: These failures are concerned with inter-module communication, and many of them are usually of a transient nature. It is to be noted that around two-thirds of errors in communication lines can be detected with the use of "vertical parity" logic.
- **Peripheral device failures**: These failures are quite important but they rarely lead to a system shutdown. The commonly occurring errors in peripheral devices are transient or intermittent, and the peripheral devices' electromechanical nature is the usual reason for their failure.
- **Environmental and power failures**: Environmental failures take place due to causes, such as air conditioning equipment failure, fires, electromagnetic interference, and earthquakes. In the case of power failures, the causes for their occurrence are the factors, such as transient fluctuations in voltage or frequency and total power loss from the local utility company.
- **Human errors**: These errors usually occur due to operator oversights and mistakes. Operator errors often take place during starting up, running, and shutting down the computer system.
- **Mysterious failures**: These failures are never categorized properly in real-time systems because they take place unexpectedly. For example, when a normally functioning system stops operating at once without indicating any problem (i.e., software, hardware, etc.) at all, the failure is called a mysterious failure.

There are many issues concerned with computer system reliability and some of the important factors to consider are presented below [8, 13, 14].

- Modern computers consist of redundancy schemes for fault tolerance, and advances made over the years have brought various types of improvements, but there are still many practical and theoretical problems that remain to be solved.
- Failures in the area of computer systems are quite highly varied in character. For example, a part/component used in a computer system may, directly or indirectly, experience a transient fault due to its surrounding environment, or it may fail permanently.

- Computers' main components/parts are the logic elements, which have quite troublesome reliability-associated features. In many situations, it is impossible for properly determining such elements' reliability and their defects cannot be healed properly.
- Prior to the installation and production phases, it could be quite difficult for detecting errors associated with hardware design at the lowest system levels. It is quite possible that oversights in hardware design may lead to situations where operational errors due to such mistakes are impossible for distinguishing from the ones due to transient physical faults.
- Usually, the most powerful type of self-repair in computer systems is dynamic fault tolerance, but it is quite difficult to analyze. However, for certain applications it is quite important and cannot be ignored.

6.3 COMPUTER FAILURE CLASSIFICATIONS, HARDWARE AND SOFTWARE ERROR SOURCES, AND COMPUTER RELIABILITY MEASURES

Computer-related failures may be categorized under the following five classifications [15]:

- **Classification I: Hardware failures.** These failures are just like in any other piece of equipment, and they occur due to factors such as poor maintenance, unexpected environmental conditions, poor design, and defective parts.
- **Classification II: Software failures.** These failures are the result of the inability of programs for continuing processing due to erroneous logic.
- **Classification III: Specifications failures.** These failures are distinguished by their origin, i.e., defects in the system's specification, rather than in the design or execution of either software or hardware.
- **Classification IV: Malicious failures.** These failures are due to a relatively new phenomenon, i.e., the malicious introduction of programs intended for causing damage to anonymous users. Often these programs are called computer viruses.
- **Classification V: Human errors.** These errors take place due to wrong actions or lack of actions by humans involved in the process (e.g., the system's operators, builders, and designers).

There are many sources for the occurrence of hardware and software errors. Some of these sources are inherited errors, data preparation errors, handwriting errors, keying errors, and optical character reader. In a computer-based system, the inherited errors can account for over 50% of the errors [16]. Furthermore, data preparation–associated tasks can also generate quite a significant proportion of errors. As per Bailey [16], at least 40% of all errors come from manipulating the data (i.e., data preparation) prior to writing it down or entering it into the involved computer system.

Additional information on computer failure classifications and hardware and software error sources is available in Refs. [15,16].

There are many measures used in the area of computer system reliability. They may be grouped under the following two categories [14,17]:

- **Category I**: This category contains four measures that are suitable for configurations such as standby, hybrid, and massively redundant systems. The measures are mean time to failure, system reliability, system availability, and mission time. It is to be noted that for evaluating gracefully degrading systems, these measures may not be sufficient.
- **Category II**: This category contains the following five new measures for handling gracefully degrading systems.
  - **Measure I: Mean computation before failure**: This is the expected amount of computation available on the system prior to failure.
  - **Measure II: Computation reliability**: This is the failure-free probability that the system will, without an error, execute a task of length, say x, started at time t.
  - **Measure III: Computation availability**: This is the expected computation capacity of the system at a given time t.
  - **Measure IV: Capacity threshold**: This is the time at which certain value of computation availability is reached.
  - **Measure V: Computation threshold**: This is the time at which certain value of computation reliability is reached for a task whose length is, say, x.

6.4 COMPUTER HARDWARE RELIABILITY VERSUS SOFTWARE RELIABILITY

As it is very important to have a clear comprehension of the differences between hardware and software reliability, a number of comparisons of important areas are presented in Table 6.1 [12,18,19].

6.5 FAULT MASKING

The term fault masking is used in the area of fault-tolerant computing, in the sense that a system with redundancy can tolerate a number of failures/malfunctions prior to its own failure. More clearly, the implication of the term is that some kind of problem has surfaced somewhere within the framework of a digital system, but because of design, the problem does not affect the overall operation of the system under consideration.

The best known fault masking method is probably modular redundancy and is presented in the following sections [12].

6.5.1 TRIPLE MODULAR REDUNDANCY (TMR)

In this case, three identical modules/units perform the same task simultaneously and the voter compares their outputs (i.e., the modules/units) and sides with the majority [12,20]. More clearly, the TMR system fails only when more than one module/unit fails or the voter fails. In other words, the TMR system can tolerate failure of a single module/unit. An important example of the TMR system's application is the Saturn V

TABLE 6.1
Hardware and software reliability comparisons

No.	Hardware Reliability	Software Reliability
1	Wears out	Does not wear out
2	Mean time to repair (MTTR) has significance	Mean time to repair (MTTR) has no significance
3	A hardware failure is generally due to physical effects	Software failure is caused by programming error
4	It is quite possible to repair hardware by using spare modules	It is impossible to repair software failures by using spare modules
5	The hardware reliability field is quite well developed, particularly in regard to electronics	The software reliability field is relatively new
6	Obtaining good failure-associated data is a problem	Obtaining good failure-associated data is a problem
7	Hardware reliability has well-developed theory and mathematical concepts	Software reliability still lacks well-developed theory and mathematical concepts
8	Generally redundancy is effective	Redundancy may not be effective
9	Preventive maintenance is conducted to inhibit failures	Preventive maintenance has no meaning in software
10	Many hardware items fail as per the bathtub hazard rate curve	Software does not fail as per the bathtub hazard rate curve
11	The failed item/system is repaired by conducting corrective maintenance	Corrective maintenance is basically redesign
12	Interfaces are visual	Interfaces are conceptual

launch vehicle computer [12,20]. The vehicle computer used TMR with voters in the central processor and duplication in the main memory [12,21].

The block diagram of the TMR scheme is shown in Figure 6.1 and the blocks in the diagram denote modules/units and the circle voter.

For independently failing modules/units and the voter, the reliability of the system in Figure 6.1 is given by [12]

$$R_{tmv} = \left(3R^2 - 2R^3\right)R_v \tag{6.1}$$

where

R_{tmv} is the reliability of the TMR system with voter.
R is the reliability of the module/unit.
R_v is the reliability of the voter.

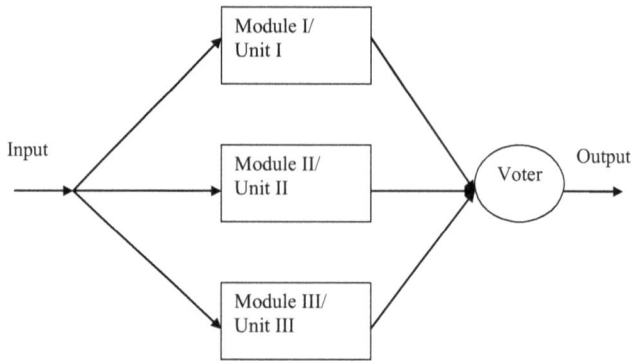

FIGURE 6.1 Block diagram for TMR system with voter.

With a perfect voter (i.e., 100% reliable), Equation (6.1) becomes

$$R_{tmp} = 3R^2 - 2R^3 \tag{6.2}$$

where

R_{tmp} is the reliability of the TMR system with perfect voter.

It is to be noted that the voter reliability and the single unit's reliability determine the improvement in reliability of the TMR system over a single unit system. For the perfect voter (i.e., $R_v = 1$), the TMR system reliability given by Equation (6.2) is only better than the single unit system when the reliability of the single unit is greater than 0.5.

At $R_v = 0.8$, the TMR system's reliability is always less than the single unit's reliability. Furthermore, when the voter reliability is 0.9 (i.e., $R_v = 0.9$), the TMR system's reliability is only marginally better than the single unit/module reliability when the single unit/module reliability is approximately between 0.667 and 0.833 [22].

6.5.1.1 TMR System Maximum Reliability with Perfect Voter

For perfect voter, the TMR system reliability is expressed by Equation (6.2). Under this scenario, the ratio of R_{tmp} to a single unit reliability, R, is given by [23]

$$\gamma = \frac{R_{tmp}}{R} = \frac{3R^2 - 2R^3}{R} = 3R - 2R^2 \tag{6.3}$$

By differentiating Equation (6.3) with respect to R and equating it to zero, we get

$$\frac{d\gamma}{dR} = 3 - 4R = 0 \tag{6.4}$$

Thus, from Equation (6.4), we obtain $R = 0.75$. This simply means that the maximum values of the reliability improvement ratio, γ, and the reliability of the TMR system, R_{tmp}, are respectively:

$$\gamma = 3(0.75) - 2(0.75)^2$$

$$= 1.125$$

and

$$R_{tmp} = 3(0.75)^2 - 2(0.75)^3$$

$$= 0.8438$$

Example 6.1

Assume that a TMR system's reliability with a perfect voter is expressed by Equation (6.2). Determine the points where the single-unit and the TMR-system reliabilities are equal.

To determine the point, we equate a single unit's reliability with Equation (6.2) to obtain

$$R = R_{tmp} = 3R^2 - 2R^3 \tag{6.5}$$

By rearranging Equation (6.5), we get

$$2R^2 - 3R + 1 = 0 \tag{6.6}$$

The above equation (i.e., Equation (6.6)) is a quadratic equation and its roots are

$$R = \frac{3 + \left[9 - 4(2)(1)\right]^{1/2}}{2(2)} = 1 \tag{6.7}$$

and

$$R = \frac{3 - \left[9 - 4(2)(1)\right]^{1/2}}{2(2)} = \frac{1}{2} \tag{6.8}$$

This means the reliabilities of the TMR system with perfect voter and the single unit are equal at $R = 1/2$ or $R = 1$. Furthermore, the reliability of the TMR system with perfect voter will only be greater than the single unit's reliability when the value of R is higher than 0.5.

6.5.1.2 TMR System with Voter Time-Dependent Reliability and Mean Time to Failure

With the aid of material presented in Chapter 3 and Equation (6.1), for constant failure rates of the TMR system units and the voter unit, the TMR system with voter reliability is expressed by [12,24].

$$R_{tmv}(t) = \left[3e^{-2\lambda t} - 2e^{-3\lambda t} \right] e^{-\lambda_{vr} t}$$

$$= 3e^{-(2\lambda + \lambda_{vr})t} - 2e^{-(3\lambda + \lambda_{vr})t} \tag{6.9}$$

where

$R_{tmv}(t)$ is the TMR system with voter reliability at time t.
λ is the unit/module constant failure rate.
λ_{vr} is the voter unit constant failure rate.

By integrating Equation (6.9) over the time interval from 0 to ∞, we get the following equation for the TMR system with voter mean time to failure [12,14]:

$$MTTF_{tmv} = \int_0^\infty \left[3e^{(2\lambda + \lambda_{vr})t} - 2e^{-(3\lambda + \lambda_{vr})t} \right] dt$$

$$= \frac{3}{(2\lambda + \lambda_{vr})} - \frac{2}{(3\lambda + \lambda_{vr})} \tag{6.10}$$

where

$MTTF_{tmv}$ is the mean time to failure of the TMR system with voter.

For perfect voter (i.e., $\lambda_{vr} = 0$), Equation (6.10) reduces to

$$MTTF_{tmp} = \frac{3}{2\lambda} - \frac{2}{3\lambda}$$

$$= \frac{5}{6\lambda} \tag{6.11}$$

where

$MTTF_{tmp}$ is the TMR system with perfect voter mean time to failure.

Example 6.2

Assume that the constant failure rate of a unit/module belonging to a TMR system with voter is $\lambda = 0.0004$ failures per hour. Calculate the system reliability for a 500-hour mission if the voter unit constant failure rate is $\lambda_{vr} = 0.0002$ failures per hour. In addition, calculate the TMR system mean time to failure.

By substituting the specified data values into Equation (6.9), we get

$$R_{tmv}(500) = 3e^{-\left[2(0.0004)+0.0002\right](500)} - 2e^{-\left[3(0.0004)+0.0002\right](500)}$$

$$= 0.8264$$

Similarly, by inserting the specified data values into Equation (6.10), we get

$$MTTF_{tmv} = \frac{3}{\left[2(0.0004)+0.0002\right]} - \frac{2}{\left[3(0.0004)+0.0002\right]}$$

$$= 1571.42 \text{ hours}$$

Thus, the TMR system with voter reliability and mean time to failure are 0.8264 and 1571.42 hours, respectively.

6.5.2 N-MODULAR REDUNDANCY (NMR)

This is the general form of the TMR (i.e., it contains N identical modules/units instead of only three units).

The number N is any odd number, and the NMR system can tolerate a maximum of n modular/unit failures if the value of N is equal to $(2n + 1)$. As the voter acts in series with the N-module system, the complete system malfunctions whenever a voter unit failure occurs.

The reliability of the NMR system with independent modules/units is given by [12,25]

$$R_{nmv} = \left[\sum_{i=0}^{n}\binom{N}{i}R^{n-1}(1-R)^{i}\right]R_{v} \tag{6.12}$$

$$\binom{N}{i} = \frac{N!}{(N-i)!i!} \tag{6.13}$$

where

R_{nmv} is the reliability of NMR system with voter.
R_{v} is the voter reliability.
R is the module/unit reliability.

Finally, it is added that the time-dependent reliability analysis of an NMR system can be performed in a manner similar to the TMR system reliability analysis. Additional information on redundancy schemes is available in Nerber [26].

6.6 SOFTWARE RELIABILITY ASSESSMENT METHODS

There are many quantitative and qualitative methods that can be used for assessing software reliability. They may be grouped under the following three classifications [22]:

- Classification I: Software Metrics
- Classification II: Software Reliability Models
- Classification III: Analytical Methods

Each of the above classifications is described in the following subsections.

6.6.1 CLASSIFICATION I: SOFTWARE METRICS

These metrics may simply be described as quantitative indicators of degree to which a software item/process possesses a specified attribute. Often, software metrics are employed for determining the status of a trend in a software development process as well as for determining the risk of going from one phase to another. Two software metrics considered quite useful for assessing, directly or indirectly, software reliability are presented in the following subsections.

6.6.1.1 Design Phase Measure

This metric is concerned with determining the degree of reliability growth during the design phase. The measure/metric requires establishing necessary defect severity classifications and possesses some ambiguity, since its low value may mean either a poor review process or a good product. The metric is expressed by [12,14,22].

$$CDR_d = \left(\sum_{i=1}^{k} \theta_i \right) / \gamma \qquad (6.14)$$

where

CDR_d is the cumulative defect ratio for design.
k is the number of reviews.
θ_i is the number of unique defects at or above a stated severity level, discovered in the ith design review.
γ is the total number of source lines of design statement in the design phase, expressed in thousands.

Additional information on this metric is available in Refs. [12,14,22].

6.6.1.2 Code and Unit Test Phase Measure

This metric is concerned with assessing software reliability during the code and unit test phase and is expressed by [12,14,22]

$$CDR_c = \left(\sum_{i=1}^{k} \alpha_i \right) / \beta \qquad (6.15)$$

where

CDR_c is the cumulative defect ratio for code.

k is the number of reviews.

α_i is the number of unique defects at or above a stated severity level, discovered in the ith code review.

β is the total number of source lines of code reviewed, expressed in thousands.

6.6.2 CLASSIFICATION II: SOFTWARE RELIABILITY MODELS

There are many software reliability models and they may be grouped under four categories as shown in Figure 6.2 [12, 14, 22, 27–29].

Category I (i.e., fault seeding) includes those software reliability models that determine the number of faults in the program at zero time via seeding of extraneous faults. Two main assumptions associated with the models belonging to this category are as follows:

- Indigenous and seeded faults have same probability of detection.
- All the seeded faults are distributed randomly in the software program under consideration.

Mills model is an example of the models belonging to this category [30].

Category II (i.e., failure count) includes those software reliability models that count the number of failures/faults taking place in stated time intervals. Three main assumptions associated with the models belonging to this category are as follows:

- Independent faults discovered during non-overlapping time intervals.
- Homogenously distributed testing during intervals.
- Independent test intervals.

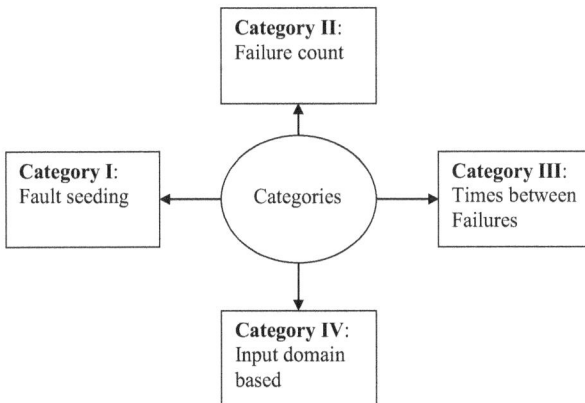

FIGURE 6.2 Categories of software reliability models.

The Musa model [31] and the Shooman model [32] are the examples of the models belonging to this category.

Category III (i.e., times between failures) includes those software reliability models that provide the time between failure estimations. Four main assumptions associated with the models belonging to this category are as follows:

- Independent times between failures.
- Independent embedded faults.
- Equal probability of exposure of each and every fault.
- Correction process does not introduce faults.

The Jelinski and Moranda model [33] and the Shick and Wolverton model [28] are the examples of the models belonging to this category.

Category IV (i.e., input domain based) includes those software reliability models that determine the software/program reliability under the condition that the test cases are sampled randomly from a given operational distribution of inputs to the software/ program. Three main assumptions associated with the models belonging to this category are as follows:

- All inputs chosen randomly.
- Input domain can be divided into equivalence groups.
- Given input profile distribution.

The Nelson model [34] and the Ramamoorthy and Bastani model [35] are the examples of the models belonging to this category.

Two, categories I and II, software reliability models are presented below.

6.6.2.1 Mills Model

The basis for this model is that an assessment of the faults remaining in a software program can be made through a seeding process that assumes a homogenous distribution of representative group of faults. Prior to the seeding process's initiation, a fault analysis is needed for determining the expected types of faults in the code as well as their relative frequency of occurrence.

An identification of seeded and unseeded faults is made during the reviews or testing process, and the discovery of indigenous and seeded faults allows an assessment of remaining faults for the type of fault under consideration. However, it is to be noted with care that the value of this measure can only be calculated if the seeded faults are discovered.

The maximum likelihood of the unseeded faults is expressed by [22, 30].

$$MLUF = (NSF)\ (NUFU)/(NSFF) \tag{6.16}$$

where

$MLUF$ is the maximum likelihood of the unseeded faults.
NSF is the number of seeded faults.

NUFU is the number of unseeded faults uncovered.
NSFF is the number of seeded faults found.

Thus, the number of unseeded faults still remaining in a software program under consideration is expressed by

$$\theta = MLUF - NUFU \tag{6.17}$$

where

θ is the number of unseeded faults still remaining in a software program under consideration.

Example 6.3

Assume that a software program was seeded with 30 faults and that, during the testing process, 60 faults of the same type were found. The breakdowns of the faults found were 35 unseeded faults and 25 seeded faults. Calculate the number of unseeded faults still remaining in the software program.

By substituting the stated data values into Equation (6.16), we get

$$MLUF = (30)(35)/25$$

$$= 42 \text{ faults}$$

By inserting the above resulting value and the stated data value into Equation (6.17), we get

$$\theta = 42 - 35$$

$$= 7 \text{ faults}$$

This means that 7 unseeded faults still remain in the software program.

6.6.2.2 Musa Model

The basis for this model is the premise that reliability assessments in the time domain can only be based upon actual execution time, as opposed to calendar/elapsed time. The main reason for this is that only during the execution process does a software program becomes exposed to failure-provoking stress.

Two of the main assumptions associated with the Musa model are as follows [12, 31]:

- Execution time between failures is piecewise exponentially distributed, and failure rate is proportional to the remaining defects.
- Failure intervals are statistically independent and follow a Poisson distribution.

The net number of corrected faults is defined by [12, 14, 31]

$$k = m\left[1 - \exp\left(-ct / mT_m\right)\right] \tag{6.18}$$

where

k is the net number of corrected faults.
m is the initial number of faults.
t is the time.
T_m is the mean time to failure at the beginning of the test.
c is the testing compression factor defined as the average ratio of detection rate of failures during test of the rate during normal use of the software program under consideration.

Mean time to failure, MTTF, increases exponentially with execution time and is expressed by

$$MTTF = T_m \exp\left(ct / mT_m\right) \tag{6.19}$$

Thus, the reliability at operational time t is

$$R(t) = \exp(-t/MTTF) \tag{6.20}$$

From the above relationships, we obtain the number of failures that must occur for improving mean time to failure from, say $MTTF_1$ to $MTTF_2$ [36]:

$$\Delta k = mT_m \left[\frac{1}{MTTF_1} - \frac{1}{MTTF_2}\right] \tag{6.21}$$

The additional execution time required to experience Δk is expressed by

$$\Delta t = \left[\frac{mT_m}{c}\right] \ln\left[\frac{MTTF_2}{MTTF_1}\right] \tag{6.22}$$

Example 6.4

Assume that a newly developed software is estimated to have approximately 400 errors. Also, at the beginning of the testing process, the recorded mean time to failure is 5 hours.

Determine the amount of time required for reducing the remaining errors to 5 if the value of the testing compression factor is 3. Also, estimate the reliability over a 150 hour operational period.

By substituting the given data values into Equation (6.21), we obtain

$$400 - 5 = (400)(5) \left[\frac{1}{5} - \frac{1}{MTTF_2} \right] \qquad (6.23)$$

Rearranging Equation (6.23) yields

$$MTTF_2 = 80 \, hours$$

By inserting the above result and the other specified data values into Equation (6.22), we get

$$\Delta t = \left[\frac{(400)(5)}{3} \right] \ln \left[\frac{80}{5} \right]$$

$$= 1848.39 \text{ hours}$$

Thus, for the given and calculated values from Equation (6.20), we obtain

$$R(150) = \exp \left(-\frac{150}{80} \right)$$

$$= 0.1533$$

Thus, the required testing time is 1848.39 hours, and the reliability of the software for the specified operational period is 0.1533.

6.6.3   CLASSIFICATION III: ANALYTICAL METHODS

There are a number of analytical methods that can be used for assessing software reliability. Two of these methods are failure modes and effect analysis (FMEA) and fault tree analysis (FTA). Both of these methods are quite commonly used for assessing reliability of hardware, and they can equally be used for assessing reliability of software as well. Both FMEA and FTA methods are described in Chapter 4.

6.7   INTERNET FACTS, FIGURES, FAILURE EXAMPLES, AND RELIABILITY-ASSOCIATED OBSERVATIONS

Some of the important Internet facts, figures, and examples are as follows:

- In 2000, in the United States Internet-related economy generated around $830 billion in revenues [9].
- From 2006 to 2011, developing countries around the globe increased their share of the world's total number of Internet users from 44% to 62% [37].
- In 2011, over 2.1 billion people around the globe were using the Internet, and approximately 45% of them were below the age of 25 years [37].
- In 2001, there were 52,658 Internet-related incidents and failures [9].

- In 2000, the Internet carried 51% of the information flowing through two-way telecommunication, and by 2007 over 97% of all telecommunicated information was transmitted through the Internet [38].
- On August 14, 1998, a misconfigured main Internet database server wrongly referred all queries for Internet systems/machines with names ending in "net" to the incorrect secondary database server. In turn, due to this problem, most of the connections to "net" Internet web servers and other end stations malfunctioned for a number of hours [8].
- On April 25, 1997, a misconfigured router of a Virginia service provider injected a wrong map into the global Internet and, in turn, the Internet providers who accepted this map automatically diverted all their traffic to the Virginia provider [39]. This caused network congestion, instability, and overload of Internet router table memory that ultimately shut down many of the main Internet backbones for around two hours [39].
- On November 8, 1998, a malformed routing control message because of a software fault triggered an inter-operability problem between a number of core Internet backbone routers produced by different vendors. In turn, this caused a widespread loss of network connectivity in addition to an increment in packet loss and latency [8]. It took many hours for most of the backbone providers for overcoming this outage.

A study in 1999 reported the following four Internet reliability–associated observations [41]:

i   Most interprovider path malfunctions occur from congestion collapse.
ii   Mean time to failure (MTTF) and mean time to repair (MTTR) for most of the Internet backbone paths are approximately 25 days or less and 20 minutes or less, respectively.
iii In the Internet backbone infrastructure, there is only a minute fraction of network paths that contribute disproportionately, directly or indirectly, to the number of long-term outages and backbone unavailability.
iv   Mean time to failure and availability of the Internet backbone structure are quite significantly lower than the Public Switched Telephone Network.

6.8   INTERNET OUTAGE CATEGORIES AND AN APPROACH FOR AUTOMATING FAULT DETECTION IN INTERNET SERVICES

A case study of Internet-related outages carried out over a period of one year has grouped the outages under 12 categories (along with their occurrences percentages in parentheses), as shown in Figure 6.3 [40].

Past experiences over the years indicate that many Internet services (e.g., e-commerce and search engines) suffer faults, and a quick detection of these faults could be an important factor in improving system availability. For this very purpose, an approach called the pinpoint method is considered extremely useful. This method combines the low-level monitors' easy deploy-ability with the higher-level monitors' ability for detecting application-level faults [41]. The method is based upon the

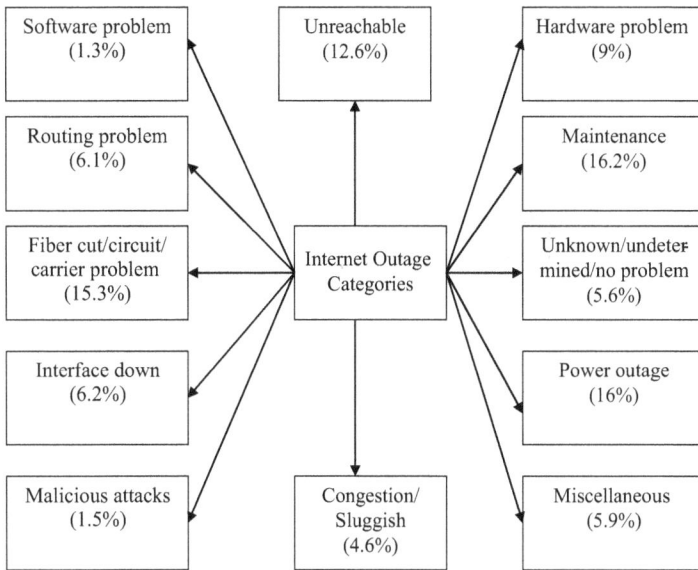

FIGURE 6.3 Categories of Internet outages (along with their occurrence percentages in parentheses).

following three assumptions with respect to the system under observation and its workload [41]:

- The software is composed of a number of interconnected modules with clearly defined narrow interfaces, which could be software subsystems, objects, or simply physical mode boundaries.
- There is a considerably higher number of basically independent requests (i.e., from different users).
- An interaction with the system is relatively short-lived, the processing of which can be decomposed as a path or, more clearly, a tree of the names of elements/parts that participate in the servicing of that request.

The pinpoint method is a three-stage process and its stages are as follows [41]:

- **Stage I: Observing the system.** This is concerned with capturing the run-time path of each and every request served/handled by the system and then, from these paths, extracting two specific low-level behaviors that are likely to reflect high-level functionality (i.e., interactions of parts/components and path shapes).
- **Stage II: Learning the patterns in system behavior.** This is concerned with constructing a reference model that clearly represents the normal behavior of an application in regard to part–component interactions and path shapes. The model is developed under the assumption that most of the system functions normally most of the time.

- **Stage III: Detecting anomalies in system behaviors.** This is basically concerned with analyzing the ongoing behaviors of the system and detecting anomalies with respect to the reference model.

Additional information on this method is available in Kiciman and Fox [41].

6.9   MODELS FOR PERFORMING INTERNET RELIABILITY AND AVAILABILITY ANALYSIS

There are many mathematical models that can be used for performing various types of reliability and availability analysis concerning the reliability of Internet-related services [12,42–45]. Two of these models are presented below.

6.9.1   MODEL I

This mathematical model is concerned with evaluating the reliability and availability of an Internet server system and it assumes that the server system can either be in an operating or a failed state. In addition, its (i.e., Internet server system) outage/failure and restoration/repair rates are constant, and all its outages/failures occur independently and the repaired/restored server system is as good as new.

The internet server system state space diagram is shown in Figure 6.4, and the numerals in rectangles represent system states.

The following symbols were used for developing equations for the model:

i is the ith Internet server system state shown in Figure 6.4 for $i = 0$ (Internet server system operating normally), $i = 1$ (Internet server system failed).
λ_{is} is the Internet server system constant outage/failure rate.
μ_{is} is the Internet server system constant restoration/repair rate.
$P_i(t)$ is the probability that the Internet server system is in state i at time t, for $i = 0, 1$.

Using the Markov method, we write down the following equations for the diagram shown in Figure 6.4 [12].

$$\frac{dP_0(t)}{dt} + \lambda_{is} P_0(t) = \mu_{is} P_1(t) \tag{6.24}$$

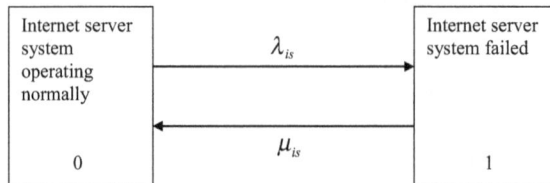

FIGURE 6.4   Internet server system state–space diagram

$$\frac{dP_1(t)}{dt} + \mu_{is} P_1(t) = \lambda_{is} P_0(t) \tag{6.25}$$

At time $t = 0$, $P_0(0) = 1$, and $P_1(0) = 0$.

By solving Equations (6.24)–(6.25), we obtain the following probability equations:

$$P_0(t) = AV_{is}(t) = \frac{\mu_{is}}{(\lambda_{is} + \mu_{is})} + \frac{\lambda_{is}}{(\lambda_{is} + \mu_{is})} e^{-(\lambda_{is} + \mu_{is})t} \tag{6.26}$$

$$P_1(t) = UA_{is}(t) = \frac{\lambda_{is}}{(\lambda_{is} + \mu_{is})} - \frac{\lambda_{is}}{(\lambda_{is} + \mu_{is})} e^{-(\lambda_{is} + \mu_{is})t} \tag{6.27}$$

where

$AV_{is}(t)$ is the Internet server system availability at time t.
$UA_{is}(t)$ is the Internet server system unavailability at time t.

As time t becomes very large, Equations (6.26) and (6.27) reduce to

$$AV_{is} = \lim_{t \to \infty} AV_{is}(t) = \frac{\mu_{is}}{(\lambda_{is} + \mu_{is})} \tag{6.28}$$

$$UA_{is} = \lim_{t \to \infty} UA_{is}(t) = \frac{\lambda_{is}}{(\lambda_{is} + \mu_{is})} \tag{6.29}$$

For $\mu_{is} = 0$, Equation (6.27) reduces to

$$R_{is}(t) = e^{-\lambda_{is} t} \tag{6.30}$$

where

$R_{is}(t)$ is the Internet server system reliability at time t.

Thus, the Internet server system mean time to failure is given by [12]

$$MTTF_{is} = \int_0^\infty R_{is}(t) dt$$

$$= \int_0^\infty e^{-\lambda_{is} t} dt$$

$$= \frac{1}{\lambda_{is}} \tag{6.31}$$

where

$MTTF_{is}$ is the Internet server system mean time to failure.

Example 6.5

Assume that the constant failure and repair rates of an Internet server system are 0.006 failures/hour and 0.05 repairs/hour, respectively. Calculate the Internet server system unavailability for a 6-hour mission.

By substituting the given data values into Equation (6.27), we get

$$UA_{is}(6) = \frac{0.006}{(0.006+0.05)} - \frac{0.06}{(0.006+0.05)} e^{-(0.006+0.05)(6)}$$

$$= 0.0305$$

Thus, the Internet server system unavailability for the stated mission time is 0.0305.

6.9.2 MODEL II

This model is concerned with evaluating the availability of an Internetworking (router) system composed of two independent and identical switches. The model assumes that the switches form a standby-type configuration and that the system fails when both the switches malfunction. Furthermore, the switch failure and restoration (repair) rates are constant. The state space diagram of the system is shown in Figure 6.5, and the numerals in boxes represent system states.

The following symbols were used for developing equations for the model:

i is the ith state shown in Figure 6.5 for: $i = 0$ (system operating normally [i.e., two switches functional: one operating, other on standby]), $i = 1$ (one switch operating, the other failed), $i = 2$ (system failed [both switches failed]).

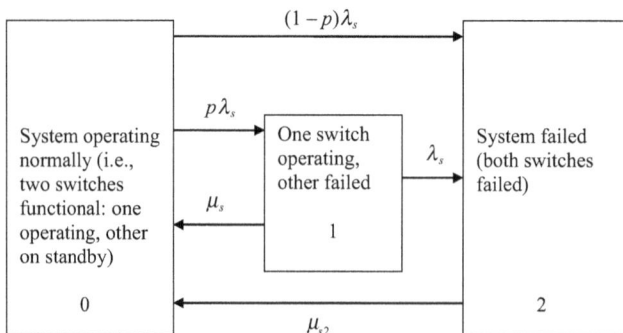

FIGURE 6.5 System state–space diagram.

p is the probability of failure detection and successful switchover from switch failure.

λ_s is the switch constant failure rate.

μ_s is the switch constant restoration/repair rate.

μ_{s2} is the constant restoration/repair rate from state 2 to state 0.

$P_i(t)$ is the probability that the Internetworking (router) system is in state *i* at time *t*; for *i* = 0, 1, 2.

Using the Markov method, we write down the equations for the diagram shown in Figure 6.5 [12,46]:

$$\frac{dP_0(t)}{dt} + \left[p\lambda_s + (1-p)\lambda_s\right]P_0(t) = \mu_s P_1(t) + \mu_{s2}P_2(t) \tag{6.32}$$

$$\frac{dP_1(t)}{dt} + (\lambda_s + \mu_s)P_1(t) = p\lambda_s P_0(t) \tag{6.33}$$

$$\frac{dP_2(t)}{dt} + \mu_{s2}P_2(t) = \lambda_s P_1(t) + (1-p)\lambda_s P_0(t) \tag{6.34}$$

At time *t* = 0, $P_0(0) = 1, P_1(0) = 0$, and $P_2(0) = 0$.

The following steady-state probability solutions are obtained by setting derivatives equal to zero in Equations (6.32)–(6.34) and using the relationship $\sum_{i=0}^{2} P_i = 1$:

$$P_0 = \mu_{s2}(\mu_s + \lambda_s)/A \tag{6.35}$$

where

$$A = \mu_{s2}(\mu_s + p\lambda_s + \lambda_s) + (1-p)\lambda_s(\mu_s + \lambda_s) + p\lambda_s^2 \tag{6.36}$$

$$P_1 = p\lambda_s\mu_{s2}/A \tag{6.37}$$

$$P_2 = \left[p\lambda_s^2 + (1-p)\lambda_s(\mu_s + \lambda_s)\right]/A \tag{6.38}$$

where

P_i is the steady-state probability that the Internetworking (router) system is in state *i*, for *i* = 0, 1, 2.

The Internetworking (router) system steady-state availability is given by

$$AV_{iss} = P_0 + P_1$$

$$= \left[\mu_{s2}(\mu_s + \lambda_s) + p\lambda_s\mu_{s2}\right]/A \tag{6.39}$$

where

AV_{iss} is the Internetworking (router) system steady-state availability.

6.10 PROBLEMS

1. What are the main causes of computer system failures?
2. Discuss at least five classifications of computer failures.
3. What are the sources of computer hardware and software errors?
4. Make a comparison between hardware and software reliability.
5. What is fault masking?
6. Compare the Mills model with the Musa model.
7. List at least ten Internet outage categories.
8. Describe the pinpoint method.
9. Prove Equations (6.26) and (6.27) by using Equations (6.24) and (6.25).
10. Prove Equation (6.39) by using Equations (6.32), (6.33), and (6.34).

REFERENCES

1. Shannon, C.E., A mathematical theory of communications, *The Bell System Technical Journal*, Vol. 27, 1948, pp. 379–423, 623–656.
2. Hamming, W.R., Error detecting and correcting codes, *The Bell System Technical Journal.*, Vol. 29, 1950, pp. 147–160.
3. Moore, E.F., Shannon, C.E., Reliable circuits using less reliable relays, *Journal of the Franklin Institute*, Vol. 262, 1956, pp. 191–208.
4. Von Neumann, J., Probabilistic Logics and the Synthesis of Reliable Organisms from Reliable Components, in *Automata Studies*, edited by C.E. Shannon and J. McCarthy, Princeton University Press, Princeton, New Jersey, 1956, pp. 43–98.
5. Haugk, G., Tsiang, S.H., Zimmerman, L., System testing of the no. 1 electronic switching system, *The Bell System Technical Journal.*, Vol. 43, 1964, pp. 2575–2592.
6. Barlow, R., Scheuer, E.M., Reliability growth during a development testing program, *Technometrics*, Vol. 8, 1966, pp. 53–60.
7. Sauter, J.L., Reliability in computer programs, *Mechanical Engineering*, Vol. 91, 1969, pp. 24–27.
8. Dhillon, B.S., *Computer System Reliability: Safety and Usability*, CRC Press, Boca Raton, Florida, 2013.
9. Goseva-Popstojanova, K., Mazidar, S., Singh, A.D., Empirical Study of Session-based Workload and Reliability for Web Servers, *Proceedings of the 15th International Symposium on Software Reliability Engineering*, 2004, pp. 403–414.
10. Yourdon, E., The causes of system failures: part II, *Modern Data*, Vol. 5, Feb. 1972, pp. 50–56.
11. Yourdon, E., The causes of system failures: part III, *Modern Data*, Vol. 5, March 1972, pp. 36–40.
12. Dhillon, B.S., *Design Reliability: Fundamentals and Applications*, CRC Press, Boca Raton, Florida, 1999.
13. Goldberg, J., *A Survey of the Design and Analysis of Fault-Tolerant Computers, in Reliability and Fault Tree Analysis*, edited by R.E. Barlow, J.B. Fussell, N.D. Singpurwalla, Society for Industrial and Applied Mathematics, Philadelphia, Pennsylvania, 1975, pp. 667–685.

14. Dhillon, B.S., *Reliability in Computer System Design*, Ablex Publishing, Norwood, New Jersey, 1987.
15. Kletz, T., Chung, P., Broomfield, E., Shen-Orr, C., *Computer Control and Human Error*, Gulf Publishing, Houston, Texas, 1995.
16. Bailey, R.W., *Human Error in Computer Systems*, Prentice Hall, Englewood Cliffs, New Jersey, 1983.
17. Beaudry, M.D., Performance related reliability measures for computer systems, *IEEE Transactions on Computers*, Vol. 27, June 1978, pp. 540–547.
18. Kline, M.B., Software and Hardware Reliability and Maintainability: What Area the Differences? *Proceedings of the Annual Reliability and Maintainability Symposium*, 1980, pp. 179–185.
19. Ireson, W.G., Coombs, C.F., Moss, R.Y., *Handbook of Reliability Engineering and Management*, McGraw Hill, New York, 1996.
20. Von Neumann, J., *Probabilistic Logics and the Synthesis of Reliable Organisms from Reliable Components, in Automata Studies*, edited by C.E. Shannon and J. McCarthy, Princeton University, Princeton, New Jersey, 1956, pp. 43–48.
21. Mathur, F.P., Avizienis, A., Reliability analysis and architecture of a hybrid redundant digital system: generalized triple modular redundancy with self- repair, *Proceedings of the AFIPS Conference*, 1970, pp. 375–387.
22. Pecht, M., Editor, *Product Reliability, Maintainability, Supportability Handbook*, CRC Press, Boca Raton, Florida, 1995.
23. Shooman, M.L., Fault-Tolerant Computing, *Annual Reliability and Maintainability Symposium Tutorial Notes*, 1994, pp. 1–25.
24. Dunn, R., Ullman, R., *Quality Assurance for Computer Software*, McGraw Hill, New York, 1982.
25. Shooman, M.L., *Reliability of Computer Systems and Networks: Fault Tolerance, Analysis, and Design*, John Wiley, New York, 2002.
26. Nerber, P.O., Power-off time impact on reliability estimates, *IEEE International Convention Record*, Part 10, March 1965, pp. 1–8.
27. Sukert, A.N., An investigation of software reliability models, *Proceedings of the Annual Reliability and Maintainability Symposium*, 1977, pp. 478–484.
28. Shick, G.J., Wolverton, R.W., An analysis of competing software reliability models, *IEEE Transactions on Software Engineering*, Vol. 4, 1978, pp. 104–120.
29. Musa, J.D., Iannino, A., Okumoto, K., *Software Reliability*, McGraw Hill, New York, 1987.
30. Mills, H.D., On the Statistical Validation of Computer Programs, Report No. 72-6015, IBM: Federal Systems Division, Gaithersburg, MD, 1972.
31. Musa, J.D., A theory of software reliability and its applications, *IEEE Transactions on Software Engineering*, Vol. 1, 1975, pp. 312–327.
32. Shooman, M.L., Software reliability measurement and models, *Proceedings of the Annual Reliability and Maintainability Symposium*, 1975, pp. 312–327.
33. Jelinski, Z., Moranda, R.B., *Software Reliability Research, In Proceedings of the Statistical Methods for the Evaluation of Computer System Performance*, Academic Press, New York, 1972, pp. 465–484.
34. Nelson, E., Estimating software reliability from test data, *Microelectronics and Reliability*, Vol. 17, 1978, pp. 67–75.
35. Ramamoorthy, C.V., Bastani, F. B., Software reliability: status and perspectives, *IEEE Transactions on Software Engineering*, Vol. 8, 1982, pp. 354–371.
36. Dunn, R., Ullman, R., *Quality Assurance for Computer Software*, McGraw Hill, New York, 1982.

37. ICT Facts and Figures, *International Telecommunication Union, ICT Data and Statistics Division*, Telecommunication Development Bureau, Geneva, Switzerland, 2011.
38. Hilbert, M., Lopez, P., The world's technological capacity to store, communicate, and compute information, *Science*, Vol. 332 (6025), April 2011, pp. 60–65.
39. Barrett, R., Haar, S., Whitestone, R., Routing snafu causes internet outage, *Interactive Week*, April 25, 1997, pp. 9.
40. Lapovitz, C., Ahuja, A., Jahamian, F., Experimental Study of Internet Stability and Wide-area Backbone Failures, *Proceedings of the 29th Annual International Symposium on Fault-Tolerant Computing*, 1999, pp. 278–285.
41. Kiciman, E., Fox, A., Detecting application-level failures in component-based internet services, *IEEE Transactions on Neural Networks*, Vol. 16, No. 5, 2005, pp. 1027–1041.
42. Hecht, M., Reliability/availability modeling and prediction of e-commerce and other internet information systems, *Proceedings of the Annual Reliability and Maintainability Symposium*, 2001, pp. 176–182.
43. Chan, C.K., Tortorella, M., Spare-Inventory Sizing for End-to-End Service Availability, *Proceedings of the Annual Reliability and Maintainability Symposium*, 2001, pp. 98–102.
44. Imaizumi, M., Kimura, M., Yasui, K., Optimal Monitoring Policy for Server System with Illegal Access, *Proceedings of the 11th ISSAT International Conference on Reliability and Quality in Design*, 2005, pp. 155–159.
45. Aida, M., Abe, T., Stochastic model of internet access patterns, *IEICE Transactions on Communications*, E84-B(8), 2001, pp. 2142–2150.
46. Dhillon, B.S., Kirmizi, F., Probabilistic safety analysis of maintainable systems, *Journal of Quality in Maintenance Engineering*, Vol. 9, No. 3, 2003, pp. 303–320.

7 Transportation Systems Failures and Reliability Modeling

7.1 INTRODUCTION

Each year, a vast sum of money is spent around the globe to develop, manufacture, and operate transportation systems such as motor vehicles, aircraft, ships, and trains. These systems carry billions of passengers and billions of tons of goods from one point to another worldwide. For example, as per the International Air Transportation Association, over 40% of world trade of goods is carried by air, and the world's airlines carry over 1.6 billion passengers each year for business and leisure travel [1].

Needless to say, transportation system failures have become a very important issue, because they can, directly or indirectly, impact the world economy, the environment, and transportation reliability.

In the area of engineering, mathematical modeling is a commonly used approach for performing various types of analysis. In this case, an item's components are denoted by idealized elements assumed to have all the representative characteristics of real-life components, and whose behavior can be described by mathematical equations. However, it is to be noted that the degree of realism of a mathematical model depends on the type of assumptions imposed upon it. Over the years, a large number of mathematical models have been developed for performing various types of engineering systems' reliability-related analyses.

This chapter presents various important aspects of motor vehicle, aircraft, ship, and train failures and mathematical models considered useful, directly or indirectly, in performing various types of transportation system reliability-related analyses.

7.2 DEFECTS IN VEHICLE PARTS AND CLASSIFICATIONS OF VEHICLE FAILURES

A motor vehicle is composed of many parts and subsystems such as steering, engine, clutch, brakes, rim, and transmission [2]. The malfunction of parts and subsystems such as these can lead to motor vehicle failure. Defects in selective automobile parts are described below [2, 3].

- **Brake defects.** In ordinary driving environments, the failure of parts in the motor vehicle braking system is quite likely to take place only when the parts become degraded, severely worn, or defective. Brake defects may be grouped under the following four categories:

    i   **Disk brake system defects.** These defects include excessive wear of the pad, excessive brake pedal travel, and low or no brake force.

    ii   **Drum brake system defects.** Some of the defects belonging to this category are brake imbalance, noise generation during braking, brake jams, low breaking performance and hard pedal, increasing heat in the brakes while driving the vehicle, and brake pedal touching the floor.

    iii **Air brake system defects.** These defects include slow brake response or release, slow pressure build-up in the reservoir, and no or low brake force.

    iv **Common disk and drum brake systems defects.** These defects include excessive pedal force, brake pedal vibrations, soft pedal, and brake fade.

- **Steering system defects.** These defects can lead to severe motor vehicle accidents and some of the causes for their occurrence are faulty changes made to the steering system, faulty design, faulty manufacturing, inadequate inspection, and poor maintenance.

- **Rim defects.** These defects are as important as defects in any other important part of a motor vehicle, because they can result in serious accidents. As per Limpert [2], one in 1300–2200 truck tire failures lead to an accident, and according to the findings of the United States Bureau of Motor Carrier Safety indicate that around 7%–13% of all trailers and tractors had at least one defective tire. Some of the causes of the rim defects are abusive operation, poor design, and faulty manufacturing operations.

Failures of a vehicle carrying passengers can be grouped under the following four categories [4]:

- **Category I.** In this case, the vehicle stops and it cannot be towed or pushed by adjacent vehicle, and it must wait for rescue vehicle.

- **Category II.** In this case, the vehicle stops or required to stop and is pushed or towed by adjacent vehicle to the close-by station. At this point, all personnel in both the affected vehicles egress, and the failed vehicle is pushed or towed for maintenance.

- **Category III.** In this case, the vehicle is required to reduce speed but is allowed to continue to the closest station, where all of its passengers must egress, and then it is dispatched for maintenance.

- **Category IV.** In this case, the vehicle is allowed to continue to the nearest station, where all its passengers must egress, and then it is dispatched for maintenance.

7.3   MECHANICAL FAILURE-ASSOCIATED AVIATION ACCIDENTS

Over the years, there have been many aviation accidents due to mechanical failures and mechanical-associated pilot errors (a mechanical-associated pilot error is the one in which pilot error was the actual cause but brought about by some kind of

TABLE 7.1
Decade breakdowns of global fatal commercial aircraft accidents due to
mechanical failure and mechanical-associated pilot error, 1950–2008

Accident cause	Time period (No. of Accidents)					
	2000–2008	1990–1999	1980–1989	1970–1979	1960–1969	1950–1959
Mechanical-associated pilot error	3	4	2	4	5	7
Mechanical failure	28	21	21	23	20	21

mechanical failure). A global study of 1300 fatal accidents involving commercial aircraft (i.e., excluding helicopters and aircraft with 10 or fewer people on board), during the period 1950–2008, highlighted a total of 134 accidents due to mechanical failure and 25 accidents due to mechanical-associated pilot error [5]. These two types of accidents are out of those accidents whose cause was clearly identifiable. The decade breakdowns of these two types of accidents are presented in Table 7.1 [5].

Eight of the aviation accidents that occurred due to mechanical failure are briefly described below.

i   **US Air Flight 427 Accident.** This accident occurred on September 8, 1994, and is concerned with the US Air Flight 427 (aircraft type: Boeing 737-387), a scheduled flight from Chicago's O'Hare Airport to West Palm Beach, Florida, via Pittsburgh, Pennsylvania [6]. The flight crashed due to rudder device malfunction and resulted in 132 fatalities.

ii   **United Airlines Flight 585 Accident.** This accident occurred on March 3, 1991, and is concerned with the United Airlines Flight 585 (aircraft type: Boeing 737-291), a scheduled flight from Stapleton International Airport, Denver, Colorado, to Colorado Springs, Colorado [7]. The flight crashed due to rudder device malfunction and resulted in 25 fatalities.

iii   **1986 British International Helicopters Chinook Accident.** This accident occurred on November 6, 1986, and is concerned with a Boeing 234LR Chinook helicopter operated by British International Helicopters [8]. The helicopter on approach to land at Sumburgh Airport, Shetland Islands, UK, crashed into the sea and sank because of the malfunction of a modified level ring gear in the forward transmission. The accident resulted in 2 injuries and 45 fatalities.

iv   **Turkish Airlines Flight 981 Accident.** This accident occurred on March 3, 1974, and is concerned with the Turkish Airlines Flight 981 (aircraft type: McDonnell Douglas DC-10-10), a scheduled flight from Istanbul, Turkey, to Heathrow Airport, London, UK, via Paris, France [9]. The flight crashed due to cargo hatch malfunction and control cable malfunctions and resulted in 346 fatalities.

v **Los Angeles Airways Flight 841 Accident.** This accident occurred on May 22, 1968, and is concerned with the Los Angeles Airways Flight 841 (aircraft type: Sikorsky S-61L helicopter), a scheduled flight from Disneyland Heliport, Anaheim, CA, to Los Angeles International Airport [10]. The flight crashed due to a mechanical failure in the blade rotor system and resulted in 23 fatalities.

vi **United Airlines Flight 859 Accident.** This accident occurred on July 11, 1961, and is concerned with the United Airlines Flight 859 (aircraft type: Douglas DC-8-20), a scheduled flight from Omaha, Nebraska, to Stapleton International Airport, Denver, Colorado [11]. The flight crashed during landing at the Stapleton International Airport because the aircraft suffered a hydraulic malfunction while en route and resulted in 84 injuries and 18 fatalities.

vii **Pan Am Flight 6 Accident.** This accident occurred on October 16, 1956, and is concerned with the Pan American World Airways (Pan Am) Flight 6 (aircraft type: Boeing 377 Stratocruiser), a scheduled flight from Honolulu, Hawaii, to San Francisco, CA [12]. The Flight was forced to ditch in the Pacific Ocean because of the malfunction of two of its four engines. Fortunately, all personnel on board survived.

viii **British Overseas Airways Corporation (BOAC) Flight 781 Accident.** This accident occurred on January 1, 1954, and is concerned with the BOAC Flight 781 (aircraft type: de Havilland DH-106 Comet 1), a scheduled flight from Kallang Airport, Singapore, to Heathrow airport, London, UK, via Rome, Italy [13]. The flight crashed into the Mediterranean Sea due to malfunction of the cabin pressure and resulted in 35 fatalities. More specifically, the failure occurred because of metal fatigue due to the repeated pressurization and depressurization of the aircraft cabin.

7.4 RAIL DEFECTS AND WELD FAILURES AND MECHANICAL FAILURE-ASSOCIATED DELAYS IN COMMUTER RAIL SERVICE

Although the rails' wear resistance is basically controlled by hardness, it is also very much dependent on the stresses that the rails are subjected to. These stresses control the defects' development in rails that can eventually lead to failure, and include contact stresses, thermal stresses, residual stresses, and bending stresses [14,15].

The contact stresses originate from the traction, wheel load, and braking and steering actions. The thermal stresses originate from welding processes during the connection of rail sections for creating a continuously welded rail, whereas the residual stresses originate from manufacturing processes. Finally, the bending stresses act either laterally or vertically, the vertical ones are mainly compressive in the rail and tensile in the rail base.

Defects in steel rails may be grouped under the following three classifications [15]:

- **Classification I: Decrease in the metal's resistance to fatigue-related defects.** These defects include the most common rail defects such as head checks and squats.

- **Classification II: Inappropriate handling, installation, and use-related defects**. These defects originate from out-of-specifications installation of rails, unexpected scratches, and wheel burns.
- **Classification III: Manufacturing defects.** These defects originate from the rail manufacturing process.

For reducing the rail-defect-related failures' occurrence, various methods are used to detect rail defects. These methods include electromagnetic acoustic transducers, visual inspection by the track maintenance staff, eddy-current testing, impedance spectroscopy, ultrasonic defect detection, and ground-penetrating radar [14,15]. Additional information on these methods is available in Refs. [14,15].

In railway systems, the construction of continuous welded rails (WR) is indispensable in improving the quality of the ride, reducing track maintenance's cost, and reducing noise and vibration. Over the years, many railway accidents have occurred due to rail weld-related failures. Thus, it is extremely important to have highly reliable welds in order to eliminate the occurrence of weld failures in service as well as for extending the CWR service life.

Past experiences over the years clearly indicate that most rail weld failures are initiated from weld discontinuities, and fusion-related welding tends to easily cause such weld discontinuities [16]. Thus, fusion-welding methods such as aluminothermic welding (TW) and enclosed-arc welding (EAW) are less reliable than pressure-welding methods such as gas-pressure welding (GPW) and flash welding (FW) [16,17]. In order to eliminate the occurrence of rail weld failures, it is highly important to conduct reliable welding by using appropriate welding processes, welding conditions, well-trained welding technicians, and adequate inspection techniques.

In many North American cities, commuter rails are used. Many times for delays in commuter rail service, mechanical failure is the main cause. In fact, as per Nelson and O'Neil [18], about one-tenth of all commuter rail service delay minutes relate to mechanical causes such as wheel defects control system problems, prime-mover failures, coach problems, hotel and head-end power failures, electrical distribution problems, and traction motor defects.

The distributions of percentages of delays due to commuter rail onboard equipment failures are as follows [18]:

- 2%: Delays caused by problems related to electrical connections.
- 2%: Delays caused by traction-motor-related problems.
- 2%: Delays caused by control-system-related problems.
- 5%: Delays caused by wheel-and axle-related problems.
- 7%: Delays caused by power systems (hotel power) for passenger lighting and for problems related to ventilation, air conditioning, and heating.
- 7%: Delays caused by problems related to coach components, such as windows, doors, toilets, and wipers in control cabs.
- 13%: Delays caused by braking system problems.
- 20%: Delays caused by prime-mover (i.e., main engine) problems.
- 42%: Delays caused by unspecified problems.

7.5   ROAD AND RAIL TANKER FAILURE MODES AND FAILURE CONSEQUENCES

Road and rail tankers are used for carrying liquefied gases and other hazardous liquids from one point to another point. Over the years, the failure of such tankers has resulted in serious consequences. The main locations of failures are valves, shells, inspection covers, pumps, connections to a container, and branches, including instrument connections.

Road and rail tanker failure modes may be grouped under three main categories, as shown in Figure 7.1 [19].

The main causes for the occurrence of metallurgical failures include erosion, fatigue, corrosion (internal or external), use of wrong or inadequate materials of construction, vessel used for purpose not covered by specification, failure to meet specified construction codes, vessel designed/constructed to an inadequate specification, and embrittlement by chemical action.

The main causes for failures due to excess internal pressure are tanker contents having higher vapor pressure than designed for, abnormal meteorological conditions, flame impingement, internal chemical reaction such as decomposition or polymerization, and hydraulic rupture consequent upon overfilling. Finally, the main causes for mechanical failures other than overpressure include collision with another vehicle, collision with a fixed object such as a bridge, general wear and tear, damage by an external explosion, modifications in violation of original specifications, and collapse of a structure onto it.

There are various consequences of road and rail tanker failures involving loss of containment. Five principal factors that influence the nature of these consequences are as follows [19]:

- **Factor I:** The location and size of any leak which develops
- **Factor II:** The physical state of the contents
- **Factor III:** The nature of the surroundings

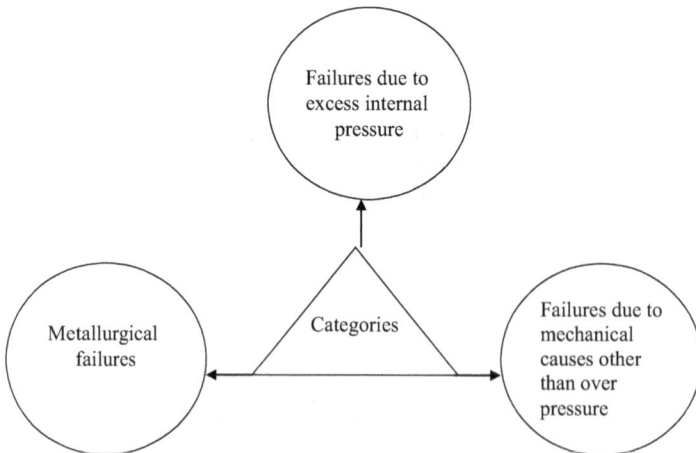

FIGURE 7.1   Road and rail tanker failure modes main categories.

- **Factor IV:** The chemical nature of the contents
- **Factor V:** The mechanism of dispersion

Additional information on these five principal factors is available in Marshall [19].

7.6   SHIP-RELATED FAILURES AND THEIR CAUSES

The shipping industrial sector is made up of many types of ships such as bulk cargo ships, container ships, tankers, and carriers. These ships contain various types of systems, equipment, and components/parts that can occasionally fail. Some examples of these systems, equipment, component/part failures are as follows:

- Propulsion system failures
- Heat-exchanger failures
- Weldment failures
- Pump failures
- Sensor failures
- Boiler failures
- Piping failures
- Fuel tank failures.

The consequences of these failures can vary quite considerably. Nonetheless, there are many distinct causes of ship failures' occurrence. Some of the common causes are shown in Figure 7.2.

7.7   FAILURES IN MARINE ENVIRONMENTS AND MICROANALYSIS TECHNIQUES FOR FAILURE INVESTIGATION

Malfunctioning of systems, equipment, or parts/components operating in marine environments can have catastrophic effects. Nonetheless, before ships sink or lie

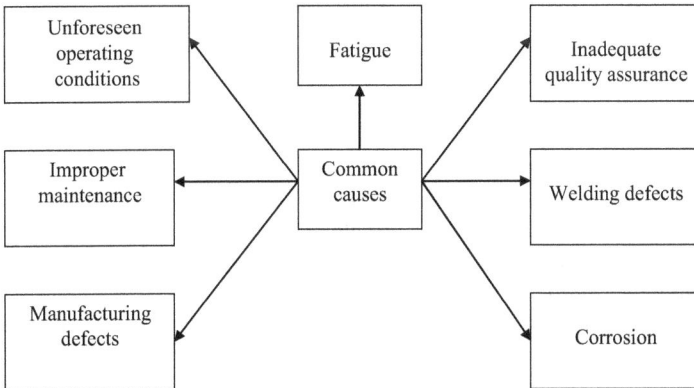

FIGURE 7.2   Ship failures' common causes.

dead in the water, a process generally occurs that causes the systems, equipment, or parts/components to breakdown. The failure mechanism may be electrical, mechanical, thermal, or chemical [20].

An electrical failure, for example, could occur as the result of internal partial discharges that degraded the insulation of a ship's propulsion motor. A mechanical failure could occur as the result of an impact between a ship and another moving vessel or a stationary object. A thermal failure could be the result of heat produced by current flowing in an electrical conductor, causing insulation degradation. Finally, a chemical failure could occur as the result of poorly protected parts'/components' corrosion on an offshore wind turbine.

Nowadays, modern vessel contains many polymeric components/parts, such as pressure seals and electrical insulation, and some of these are very critical to the vessel operation. There are many microanalysis techniques that are considered quite useful in failure investigations involving polymers. Four commonly used micro-analysis techniques are described in the following sections [20].

7.7.1 THERMOMECHANICAL ANALYSIS

This technique involves measuring variations in a sample's volume or length as a function of time or/and temperature. The technique is quite commonly used for determining thermal expansion coefficients as well as the glass-transition temperature of polymer or composite materials. A weighted probe is placed on the specimen surface, and the vertical movement of the probe is monitored on continuous basis while the sample is heated at a controlled rate.

7.7.2 THERMOGRAVIMETRIC ANALYSIS

This technique measures variations in the weight of a sample under consideration as a function of temperature or time. The technique is used for determining polymer degradation temperatures, levels of residual solvent, the degree of inorganic (i.e., non-combustible) filler in polymer or composite material compositions, and absorbed moisture content.

Finally, it is to be noted that the technique can also be quite useful in deformulation of complex polymer-based products.

7.7.3 DIFFERENTIAL SCANNING CALORIMETRY

This technique measures heat flow to a polymer. This is very important because, by monitoring the heat flow as a function of temperature, phase transitions such as glass-transition temperatures and crystalline melt temperatures can be characterized quite effectively. This, in turn, is very useful for determining how a polymer will behave at operational temperatures.

The technique can also be utilized in forensic investigations for determining the maximum temperature that a polymer has been subjected to. This can be quite useful in establishing whether an equipment/system/component has been subjected to thermal overloads during service. Finally, this technique can also be employed for determining the thermal stability of polymers by measuring the oxidation induction temperature/time.

7.7.4 Fourier Transform Infrared Spectroscopy

This technique is used for identifying and characterizing polymer materials and their additives. This is an extremely useful method, particularly in highlighting defects or inclusions in plastic films or molded parts. Additional information on this method is available in Dean [20].

7.8 MATHEMATICAL MODELS FOR PERFORMING RELIABILITY ANALYSIS OF TRANSPORTATION SYSTEMS

Mathematical modeling is a commonly used approach for performing various types of analysis in the area of engineering. In this case, the components of an item are denoted by idealized elements assumed to have all the representative characteristics of real-life components, and whose behavior can be described by mathematical equations. However, a mathematical model's degree of realism very much depends on the type of assumptions imposed upon it.

Over the years, a large number of mathematical models have been developed for performing various types of reliability-related analysis of engineering systems. Most of these models were developed using the Markov method. This section presents four such models considered useful for performing various types of transportation system reliability-related analysis.

7.8.1 Model I

This mathematical model represents a transportation system that can fail either due to human errors or hardware failures. A typical example of such a transportation system is a truck. The failed transportation system is towed to the repair workshop for repair. The state-space diagram of the transportation system is shown in Figure 7.3. The numerals in circles and boxes denote system states.

The model is subjected to the following assumptions:

* Failure and towing rates of the transportation system are constant.
* Transportation system can fail completely either due to hardware failures or human errors.
* Failures and human errors occur independently.

The following symbols are associated with the model:

j is the jth state of the transportation system, where $j = 0$ (transportation system operating normally), $j = 1$ (transportation system failed in the field due to a hardware failure), $j = 2$ (transportation system failed in the field due to a human error), $j = 3$ (transportation system in the repair workshop).

λ_{hd} is the transportation system constant hardware failure rate.

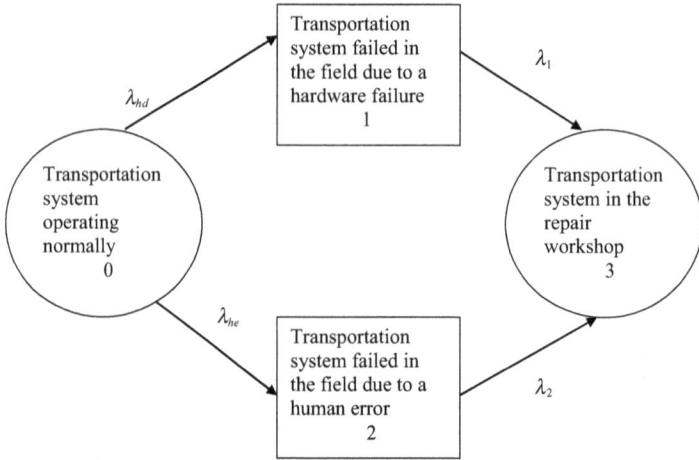

FIGURE 7.3 State-space diagram for Model I.

λ_{hr} is the transportation system constant failure rate due to human errors.
λ_1 is the transportation system constant towing rate from state 1 to state 3.
λ_2 is the transportation system constant towing rate from 2 to state 3.
$P_j(t)$ is the probability that the transportation system is in state j at time t, for $j = 0$, 1, 2, 3.

Using the Markov method presented in Chapter 4 and Figure 7.3, we write down the following equations [21–23]:

$$\frac{dP_0(t)}{dt} + \left(\lambda_{hd} + \lambda_{he}\right)P_0(t) = 0 \tag{7.1}$$

$$\frac{dP_1(t)}{dt} + \lambda_1 P_1(t) = \lambda_{hd} P_0(t) \tag{7.2}$$

$$\frac{dP_2(t)}{dt} + \lambda_2 P_2(t) = \lambda_{he} P_0(t) \tag{7.3}$$

$$\frac{dP_3(t)}{dt} = \lambda_1 P_1(t) + \lambda_2 P_2(t) \tag{7.4}$$

At $t = 0$, $P_0(0) = 1, P_1(0) = 0, P_2(0)$, and $P_3(0) = 0$.

By solving Equations (7.1)–(7.4), we get the following state probability equations [21–23]:

$$P_0(t) = e^{-at} \tag{7.5}$$

where

$$a = \lambda_{hd} + \lambda_{he} \tag{7.6}$$

$$P_1(t) = c_2 \left(e^{-at} - e^{-\lambda_1 t} \right) \tag{7.7}$$

where

$$c_2 = \frac{\lambda_{hd}}{\left(\lambda_1 - a \right)} \tag{7.8}$$

$$P_2(t) = c_1 \left(e^{-at} - e^{-\lambda_2 t} \right) \tag{7.9}$$

where

$$c_1 = \frac{\lambda_{he}}{\left(\lambda_2 - a \right)} \tag{7.10}$$

$$P_3(t) = 1 + c_1 e^{-\lambda_2 t} + c_2 e^{-\lambda_1 t} + c_3 e^{-\left(\lambda_{hd} + \lambda_{he} \right)t} \tag{7.11}$$

where

$$c_3 = -\left[c_1 \lambda_2 + c_2 \lambda_1 \right] / a \tag{7.12}$$

The transportation system reliability is given by

$$R_{ts}(t) = P_0(t) = e^{-at} = e^{-\left(\lambda_{hd} + \lambda_{he} \right)t} \tag{7.13}$$

where

$R_{ts}(t)$ is the transportation system reliability at time t.

The transportation system mean time to failure is expressed by [21–24].

$$MTTF_{ts} = \int_0^\infty R_{ts}(t) dt$$

$$= \int_0^\infty e^{-\left(\lambda_{hd} + \lambda_{he} \right)t} dt$$

$$= \frac{1}{\left(\lambda_{hd} + \lambda_{he} \right)} \tag{7.14}$$

where

$MTTF_{ts}$ is the transportation system mean time to failure.

Example 7.1

Assume that a transportation system hardware failure and failure due to human error rates are 0.0004 failures/hour and 0.0003 failures/hour, respectively. Calculate the transportation system reliability during a 10-hour mission and mean time to failure.

By substituting the given data values into Equation (7.13), we obtain

$$R_{ts}(10) = e^{-(0.0004+0.0003)(10)}$$

$$= 0.9930$$

Also, by substituting the specified data values into Equation (7.14), we obtain

$$MTTF_{ts} = \frac{1}{(0.0004+0.0003)}$$

$$= 1428.57 \text{ hours}$$

Thus, the transportation system reliability and mean time to failure are 0.9930 and 1428.57 hours, respectively.

7.8.2 MODEL II

This mathematical model represents a three-state transportation system in which a vehicle can be in any one of the three states: vehicle functioning normally in the field, vehicle failed in the field, and failed vehicle in the repair workshop. The failed vehicle is taken to the repair workshop from the field. The repaired vehicle is put back to its normal operating/functioning state. The transportation system state-space diagram is shown in Figure 7.4. The numerals in the circles and box denote transportation system states.

The model is subjected to the following assumptions:

- Vehicle failure and towing rates are constant
- Vehicle repair rate is constant
- A repaired vehicle is as good as new
- Vehicle failures occur independently.

The following symbols are associated with the model:

j is the jth state of the vehicle/transportation system, where $j = 0$ (vehicle functioning normally in the field), $j = 1$ (vehicle failed in the field), $j = 2$ (failed vehicle in the repair workshop).

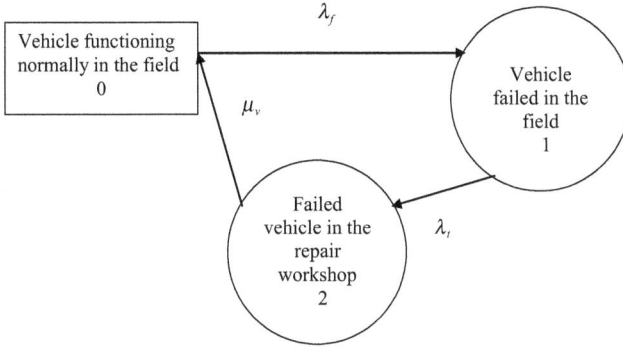

FIGURE 7.4 State-space diagram for Model II.

λ_f is the vehicle constant failure rate.
λ_t is the vehicle constant towing rate.
μ_v is the vehicle constant repair rate.
$P_j(t)$ is the probability that the vehicle/transportation system is in state j at time t, for $j = 0, 1, 2$.

Using the Markov method presented in Chapter 4 and Figure 7.4, we write down the following equations [25]:

$$\frac{dP_0(t)}{dt} + \lambda_f P_0(t) = \mu_v P_2(t) \tag{7.15}$$

$$\frac{dP_1(t)}{dt} + \lambda_t P_1(t) = \lambda_f P_0(t) \tag{7.16}$$

$$\frac{dP_2(t)}{dt} + \mu_v P_2(t) = \lambda_t P_1(t) \tag{7.17}$$

At time $t = 0$, $P_0(0) = 1, P_1(0) = 0, and\ P_2(0) = 0$.

By solving Equations (7.15)–(7.17), we obtain the following steady-state probability equations [25]:

$$P_0 = \left[1 + \frac{\lambda_f}{\lambda_t} + \frac{\lambda_f}{\mu_v} \right]^{-1} \tag{7.18}$$

$$P_1 = \left(\frac{\lambda_f}{\lambda_t} \right) P_0 \tag{7.19}$$

$$P_2 = \left(\frac{\lambda_f}{\mu_v} \right) P_0 \qquad (7.20)$$

where

$P_0, P_1, and\ P_2$ are the steady-state probabilities of the vehicle/transportation system being in states 0, 1, and 2, respectively.

The vehicle/transportation system steady-state availability is given by

$$AV_v = P_0 \qquad (7.21)$$

where

AV_v is the vehicle/transportation system steady-state availability.

By setting $\mu_v = 0$ in Equations (7.15)–(7.17) and then solving the resulting equations, we get

$$R_v(t) = P_0(t) = e^{-\lambda_f t} \qquad (7.22)$$

where

$R_v(t)$ is the vehicle/transportation system reliability at time t.
$P_0(t)$ is the probability of the vehicle/transportation system being in state 0 at time t.

The vehicle/transportation system mean time to failure is expressed by [24]

$$MTTF_v = \int_0^\infty R_v(t)dt$$

$$= \int_0^\infty e^{-\lambda_f t} dt$$

$$= \frac{1}{\lambda_f} \qquad (7.23)$$

where

$MTTF_v$ is the vehicle/transportation system mean time to failure.

Example 7.2

Assume that a three-state transportation system constant failure rate is 0.0002 failures/hour. Calculate the transportation system reliability during a 5-hour mission and mean time to failure.

By substituting the specified data values into Equation (7.22), we obtain

$$R_v(5) = e^{-(0.0002)(5)}$$

$$= 0.9990$$

Also, by substituting the given data value into Equation (7.23), we get

$$MTTF_v = \frac{1}{0.0002} = 5,000 \text{ hours}$$

Thus, the transportation system reliability and mean time to failure are 0.9990 and 5,000 hours, respectively.

7.8.3 MODEL III

This mathematical model represents a three-state transportation system in which a vehicle is functioning in alternating weather (e.g., normal and stormy). The vehicle can malfunction either in normal or stormy weather. The failed (i.e., malfunctioned) vehicle is repaired back to both its operating states. The system state-space diagram is shown in Figure 7.5. The numerals in circles and a box denote system states.

The model is subjected to the following assumptions:

- Vehicle failure and repair rates are constant
- Alternating weather transition rates (i.e., from normal weather state to stormy weather state and vice versa) are constant
- Vehicle failures occur independently
- A repaired vehicle is as good as new.

The following symbols are associated with the model:

j is the jth state of the vehicle/transportation system, where $j = 0$ (vehicle functioning in normal weather), $j = 1$ (vehicle functioning in stormy weather), $j = 2$ (vehicle failed).
λ_n is the vehicle constant failure rate for normal weather state.
λ_s is the vehicle constant failure rate for stormy weather state.
μ_1 is the vehicle constant repair rate (normal weather) from state 2 to state 0.
μ_2 is the vehicle constant repair rate (stormy weather) from state 2 to state 1.
θ is the weather constant changeover rate from state 0 to state 1.

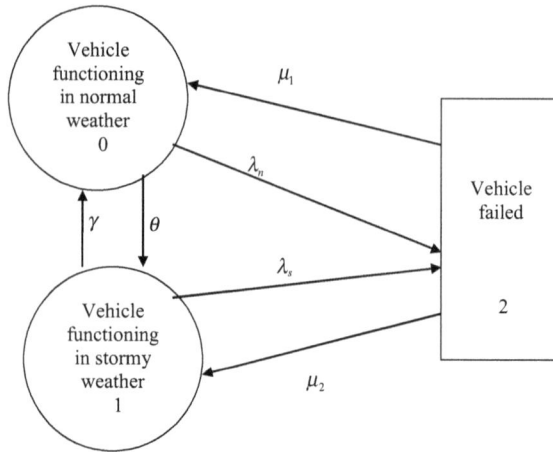

FIGURE 7.5 State-space diagram for Model III.

γ is the weather constant changeover rate from state 1 to state 0.

$P_j(t)$ is the probability that the vehicle/transportation system is in state j at time t, for $j = 0, 1, 2$.

Using Markov method presented in Chapter 4 and Figure 7.5, we write down the following equations [26]:

$$\frac{dP_0(t)}{dt} + (\theta + \lambda_n)P_0(t) = \gamma P_1(t) + \mu_1 P_2(t) \tag{7.24}$$

$$\frac{dP_1(t)}{dt} + (\gamma + \lambda_s)P_1(t) = \theta P_0(t) + \mu_2 P_2(t) \tag{7.25}$$

$$\frac{dP_2(t)}{dt} + (\mu_1 + \mu_2)P_2(t) = \lambda_n P_0(t) + \lambda_s P_1(t) \tag{7.26}$$

At time $t = 0$, $P_0(0) = 1$, $P_1(0) = 0$, *and* $P_2(0) = 0$.

By solving Equations (7.24)–(7.26), we get the following steady-state probability equations [26]:

$$P_0 = B_1 / F_1 F_2 \tag{7.27}$$

where

$$B_1 = \mu_1 \gamma + \lambda_s \mu_1 + \gamma \mu_2 \tag{7.28}$$

$$F_1, F_2 = \frac{-G \pm \left[G^2 - 4(B_1 + B_2 + B_3) \right]^{1/2}}{2} \tag{7.29}$$

$$B_2 = \theta\lambda_s + \gamma\mu_1 + \lambda_n\lambda_s \tag{7.30}$$

$$B_3 = \theta\mu_1 + \theta\mu_2 + \lambda_n\mu_2 \tag{7.31}$$

$$G = \gamma + \mu_1 + \mu_2 + \theta + \lambda_n + \lambda_s \tag{7.32}$$

$$P_1 = B_3 / F_1 F_2 \tag{7.33}$$

$$P_2 = B_2 / F_1 F_2 \tag{7.34}$$

where

$P_0, P_1, and\ P_2$ are steady-state probabilities of the vehicle/transportation system being in states 0, 1, and 2, respectively.

The vehicle steady-state availability in both types of weather is expressed by

$$VA_{ss} = P_0 + P_1 \tag{7.35}$$

where

VA_{ss} is the vehicle steady-state availability in both types of weather.

By setting $\mu_1 = \mu_2 = 0$ in Equations (7.24)–(7.26) and then solving the resulting equations [24, 26], we obtain

$$MTTF_{ve} = \lim_{s \to 0} R_{ve}(s) = \lim_{s \to 0}\left[P_0(s) + P_1(s)\right]$$

$$= \frac{\lambda_s + \theta + \gamma}{(\lambda_n + \theta)(\lambda_s + \gamma) - \theta\gamma} \tag{7.36}$$

where

$MTTF_{ve}$ is the vehicle mean time to failure.
s is the Laplace transform variable.
$R_{ve}(s)$ is the Laplace transform of the vehicle reliability.
$P_0(s)$ is the Laplace transform of the probability that the vehicle is in state 0.
$P_1(s)$ is the Laplace transform of the probability that the vehicle is in state 1.

Example 7.3

Assume that in Equation (7.36), we have the following given data values:

$\theta = 0.0004$ transitions/hour
$\gamma = 0.0005$ transitions/hour

$\lambda_n = 0.0006$ failures/hour
$\lambda_s = 0.0008$ failures/hour

Calculate mean time to failure of the vehicle.
By substituting the specified data values into Equation (7.36), we obtain

$$MTTF_{ve} = \frac{(0.0008)+(0.0004)+(0.0005)}{(0.0006+0.0004)(0.0008+0.0005)-(0.0004)(0.0005)}$$
$$= 1545.45\,\text{hours}$$

Thus, mean time to failure of the vehicle is 1545.45 hours.

7.8.4 MODEL IV

This mathematical model represents a four-state transportation system in which a transportation system can be in any one of the four states: transportation system operating normally in the field, transportation system failed safely in the field, transportation system failed with accident in the field, and failed transportation system in the repair workshop. The failed transportation system is taken to the repair workshop from the field. The repaired transportation system is put back into its normal operation.

The transportation system state-space diagram is shown in Figure 7.6. The numerals in circles and boxes denote transportation system states.

The model is subjected to the following assumptions:

- Transportation system safe failure and accident repair rates are constant.
- Transportation system towing and repair rates are constant.

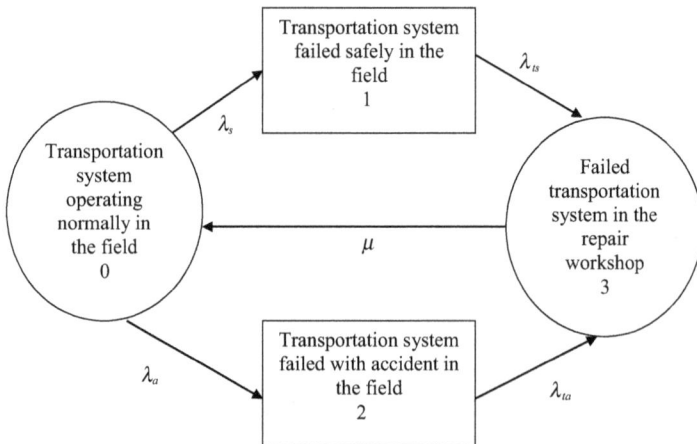

FIGURE 7.6 State-space diagram for Model IV.

- Transportation system failures occur independently.
- A repaired transportation system is as good as new.

The following symbols are associated with the model:

j is the jth state of the transportation system, where $j = 0$ (transportation system operating normally in the field), $j = 1$ (transportation system failed safely in the field), $j = 2$ (transportation system failed with accident in the field), $j = 3$ (failed transportation system in the repair workshop).

λ_s is the transportation system fail-safe constant failure rate.

λ_a is the transportation system fail-accident constant failure rate.

λ_{ts} is the transportation system constant towing rate from state 1.

λ_{ta} is the transportation system constant towing rate from state 2.

μ is the transportation system constant repair rate.

$P_j(t)$ is the probability that the transportation system is in state j at time t, for $j = 0$, 1, 2, 3.

Using the Markov method presented in Chapter 4 and Figure 7.6, we write down the following equations [25]:

$$\frac{dP_0(t)}{dt} + (\lambda_s + \lambda_a) P_0(t) = \mu P_3(t) \tag{7.37}$$

$$\frac{dP_1(t)}{dt} + \lambda_{ts} P_1(t) = \lambda_s P_0(t) \tag{7.38}$$

$$\frac{dP_2(t)}{dt} + \lambda_{ta} P_2(t) = \lambda_a P_0(t) \tag{7.39}$$

$$\frac{dP_3(t)}{dt} + \mu P_3(t) = \lambda_{ts} P_1(t) + \lambda_{ta} P_2(t) \tag{7.40}$$

At time $t = 0$, $P_0(0) = 1$, $P_1(0) = 0$, $P_2(0) = 0$, and $P_3(0) = 0$.

By solving Equations (7.37)–(7.40), we get the following steady-state probability equations [25]:

$$P_0 = \left[1 + \frac{\lambda_s}{\lambda_{ts}} + \frac{\lambda_a}{\lambda_{ta}} + \frac{(\lambda_s + \lambda_a)}{\mu} \right]^{-1} \tag{7.41}$$

$$P_1 = \frac{\lambda_s P_0}{\lambda_{ts}} \tag{7.42}$$

$$P_2 = \frac{\lambda_a P_0}{\lambda_{ta}}$$

(7.43)

$$P_3 = \frac{(\lambda_s + \lambda_a) P_0}{\mu}$$

(7.44)

where

$P_0, P_1, P_2,$ and P_3 are the steady-state probabilities of the transportation system being in states 0, 1, 2, and 3, respectively.

The transportation system steady-state availability is expressed by

$$AV_{ts} = P_0$$

(7.45)

where

AV_{ts} is the transportation system steady-state availability.

By setting $\mu = 0$ in Equations (7.37)–(7.40) and then solving the resulting equations, we obtain

$$R_{ts}(t) = e^{-(\lambda_s + \lambda_a)t}$$

(7.46)

where

$R_{ts}(t)$ is the transportation system reliability at time t.

The transportation system mean time to failure is expressed by [24]

$$MTTF_{ts} = \int_0^\infty R_{ts}(t)\, dt$$

$$= \int_0^\infty e^{-(\lambda_s + \lambda_a)t}\, dt$$

$$= \frac{1}{(\lambda_s + \lambda_a)}$$

where

$MTTF_{ts}$ is the transportation system mean time to failure.

Example 7.4

Assume that in Equation (7.41), we have the following given data values:

$\lambda_s = 0.0008$ failures/hour
$\lambda_a = 0.0004$ failures/hour
$\lambda_{ts} = 0.0007$ towings/hour
$\lambda_{ta} = 0.0004$ towings/hour
$\mu = 0.0005$ repairs/hour

Calculate the transportation system steady-state availability.
By substituting the given data values into Equation (7.41), we get

$$P_0 = AV_{ts} = \left[1 + \frac{(0.0008)}{(0.0007)} + \frac{(0.0004)}{(0.0004)} + \frac{(0.0008 + 0.0004)}{(0.0005)} \right]^{-1}$$

$$= 0.1804$$

Thus, the transportation system steady-state availability is 0.1804.

7.9 PROBLEMS

1. Discuss at least three types of defects in vehicle parts.
2. Discuss the following mechanical failure-associated aviation accidents:
   - US Air Flight 427 accident
   - United Airlines Flight 585 accident
   - British Overseas Airways Corporation (BOAC) Flight 781 accident
3. Discuss the classifications of defects in steel rails.
4. Discuss rail weld failures and mechanical failure-associated delays in commuter rail service.
5. Discuss road and rail tanker failure modes and failure consequences.
6. What are the common causes of ship failures?
7. Describe the following two techniques:
   i) Differential scanning calorimetry
   ii) Therogravimetric analysis
8. Assume that a transportation system hardware failure and failure due to human error rates are 0.0006 failures/hour and 0.0002 failures/hour, respectively. Calculate the transportation system reliability during an 8-hour mission and mean time to failure.
9. Prove Equation (7.36) by setting $\mu_1 = \mu_2 = 0$ in Equations (7.24)–(7.25) and then using the resulting equations.
10. Prove Equations (7.41)–(7.44) by using Equations (7.37)–(7.40).

REFERENCES

1. Fast Facts: The Air Transport Industry in Europe has United to Present Its Key Facts and Figures, IATA, 2006, International Air Transport Association, Montreal, Canada. www.iata.org/pressroom/economics.facts/stats/2003-04-10-01.htm, Accessed on March 20, 2008.
2. Limpert, R., *Vehicle System Components: Design and Safety*, John Wiley, New York, 1982.
3. Dhillon, B.S., *Mechanical Reliability: Theory, Models, and Applications*, American Institute of Aeronautics and Astronautics, Washington, DC, 1988.
4. Anderson, J.E., *Transit Systems Theory*, Heath Company, Lexington, Massachusetts, 1978.
5. Causes of Fatal Aviation Accidents by Decade, 2009, www.planecrashinfo.com/cause.htm, Accessed on December 10, 2010.
6. Bryne, G., *Flight 427: Anatomy of an Air Disaster*, Springer-Verlag, New York, 2002.
7. Aircraft Accident Report: United Airlines Flight 585, Report No. AAR92-06, *National Transportation Safety Board*, Washington, DC, 1992. http://libraryonline.erau.edu/online-full-text/ntsb/aircraft-accident-reports/AAR92-06.pdf, Accessed on December 15, 2010.
8. Report on the Accident to Boeing Vertol (BV) 234LR, G-BWFC 2.5 miles east of Sumburgh, Shetland Isles, 6 November 1986, Report No. 2, Air Accidents Investigation Branch (AAIB), Aldershot, UK, 1986.
9. Johnston, M., *The Last Nine Minutes: The Story of Flight 981*, Morrow, New York, 1976.
10. Gero, D., *Aviation Disasters*, Patrick Stephens Ltd., Sparkford, UK, 1993.
11. United Airlines Flight 859, *Aircraft Accident Report No. SA-362 (file 1-0003)*, Civil Aeronautics Board (CAB), Washington, DC, 1962.
12. Pan Am Flight 6 (Registration No. 90943), *Accident Investigation Report File No. 1-0121*, Department of Transportation, Washington, DC, 1957.
13. Stewart, S., *Air Disasters*, Arrow Books, London, 1986.
14. Cannon, D.F., Edel, K.O., Grassie, S.L., Sawley, K., Rail defects: an overview, *Fatigue and Fracture of Engineering Materials and Structures*, Vol. 26, No 10, 2003, pp. 865–886.
15. Labropoulos, K.C., Moundoulas, P., Moropoulou, A., Methodology for the Monitoring, Control, and Warning of Defects for Preventive Maintenance of Rails, in *Computers in Railways*, X. Vol. 88, WIT Transactions on the Built Environment, WIT Press, London, 2006, pp. 935–944.
16. Fukada, Y., Yammamotto, R., Harasawa, H., Nakanowatari, Experience in Maintaining Rail Track in Japan, *Welding in the Weld*, Vol. 47, 2003, pp. 123–137.
17. Tatsumi, M., Fukada, Y., Veyama, K., Shitara, H., Yamamoto, R., Quality evaluation methods for rail welds in Japan, *Proceedings of the World Congress on Railway Research*, 1997, pp. 197–205.
18. Nelson, D., O'Neil, K., Commuter Rail Reliability: On-Time Performance and Causes for Delays, *Transportation Research Record* 1704, 2000, pp. 42–50.
19. Marshall, V.C., Modes and Consequences of the Failure of Road and Rail Tankers Carrying Liquefied Gases and Other Hazardous Liquids, in *Reliability on the Move*, edited by G.B. Guy, Elsevier Science, London, 1989, pp. 136–148.
20. Dean, R.J., *Investigation of Failure in Marine Environment*, ERA Technology Ltd., Leatherhead, Surrey, UK, 2009.
21. Dhillon, B.S., *Human Factors: With Human Factors*, Pergamon Press, New York, 1986.

22. Dhillon, B.S., Rayapati, S.N., Reliability and Availability Analysis of on Surface Transit Systems, *Microelectronics and Reliability*, Vol. 24, 1984, pp. 1029–1033.

23. Dhillon, B.S., Rayapati, S.N., Reliability Evaluation of Transportation Systems with Human Errors, *Proceedings of the IASTED International Conference on Applied Simulation and Modeling*, 1985, pp. 4–7.

24. Dhillon, B.S., *Design Reliability: Fundamentals and Applications*, CRC Press, Boca, Raton, Florida, 1999.

25. Dhillon, B.S., Rayapati, S.N., Reliability and Availability Analysis of Transit Systems, *Microelectronics and Reliability*, Vol. 25, No. 6, 1985, pp. 1073–1085.

26. Dhillon, B.S., RAM Analysis of Vehicles in Changing Weather, *Proceedings of the Annual Reliability and Maintainability Symposium*, 1984, pp. 48–53.

8 Power System Reliability

8.1 INTRODUCTION

The main areas of an electric power system are generation, transmission, and distribution and the basic function of a modern electric power system is supplying its customers the cost-effective electric energy with a high degree of reliability. In the context of an electric power system, reliability may simply be expressed as concern regarding the system's ability for providing a satisfactory amount of electrical power [1].

The history of power system reliability goes back to the early years of 1930s when probability concepts were applied to electric power system-associated problems [2,3]. Over the years, a large number of publications on the subject have appeared. Most of these publications are listed in Refs. [4–7].

This chapter presents various important aspects of power system reliability.

8.2 POWER SYSTEM RELIABILITY-RELATED TERMS AND DEFINITIONS

There are many terms and definitions used in the area of power system reliability. Some of these are as follows [8–11]:

- **Forced outage.** This is when a unit or a piece of equipment has to be taken out of service because of damage or a component failure.
- **Forced outage rate.** This is (for equipment) expressed by the total number of forced outage hours times 100 over the total number of service hours plus the total number of forced outage hours.
- **Power system reliability.** This is the degree to which the performance of the elements in a bulk system results in electrical energy being delivered to customers within the framework of stated standards and in the amount needed.
- **Scheduled outage.** This is the shutdown of a generating unit, transmission line, or other facility, for inspection or maintenance, as per an advance schedule.
- **Forced outage hours.** These are the total number of hours a unit or a piece of equipment spends in the forced outage condition.

- **Forced derating.** This is when a unit or a piece of equipment is operated at a forced derated or lowered capacity because of a component failure or damage.
- **Mean time to forced outage.** This is analogous to mean time to failure (MTTF) and is expressed by the total of service hours over the total number of forced outages.
- **Mean forced outage duration.** This is analogous to mean time to repair (MTTR) and is expressed by the total number of forced outage hours over the total number of forced outages.

8.3   POWER SYSTEM SERVICE PERFORMANCE INDICES

In the electric power system area, generally various service performance-related indices are calculated for the total system, a specific region or voltage level, designated feeders or different groups of customers, etc. [1]. Six of these indices are as follows [1,12].

8.3.1   INDEX I

This index is known as average service availability index and is defined by

$$\theta_1 = \frac{\alpha_1}{\gamma_1} \tag{8.1}$$

where

θ_1 is the average service availability.
α_1 is the customer hours of available service.
γ_1 is the customer hours demanded. These hours are expressed by the 12- month average number of customers serviced times 8760 hours.

8.3.2   INDEX II

This index is known as system average interruption duration index and is defined by

$$\theta_2 = \frac{\alpha_2}{\gamma_2} \tag{8.2}$$

where

θ_2 is the system average interruption duration.
α_2 is the sum of customer interruption durations per year.
γ_2 is the total number of customers.

8.3.3 INDEX III

This index is known as customer average interruption duration index and is defined by

$$\theta_3 = \frac{\alpha_3}{\gamma_3} \tag{8.3}$$

where

θ_3 is the customer average interruption duration.
γ_3 is the total number of customer interruptions per year.
α_3 is the sum of customer interruption durations per year.

8.3.4 INDEX IV

This index is known as customer average interruption frequency index and is expressed by

$$\theta_4 = \frac{\alpha_4}{\gamma_4} \tag{8.4}$$

where

θ_4 is the customer average interruption frequency.
α_4 is the total number of customer interruptions per year.
γ_4 is the total number of customers affected. It is to be noted that the customers affected should only be counted once, irrespective of the number of interruptions they may have experienced throughout the year.

8.3.5 INDEX V

This index is known as system average interruption frequency index and is expressed by

$$\theta_5 = \frac{\alpha_5}{\gamma_5} \tag{8.5}$$

where

θ_5 is the system average interruption frequency.
α_5 is the total number of customer interruptions per year.
γ_5 is the total number of customers.

8.3.6 INDEX VI

This index is concerned with measuring service quality (i.e., measuring the continuity of electricity supply to the customer) and is defined by [5,12]

$$\theta_6 = (8760)(MTEI)/[(MTTF)(8760)+(MTEI)] \qquad (8.6)$$

where

θ_6 is the mean number of annual down hours (i.e., service outage hours) per customer.
(8760) is the total number of hours in one year.
MTEI is the mean time of electricity interruption.
MTTF is the mean time to failure (i.e., the average time between electricity interruptions).

Example 8.1

Assume that the annual failure rate of the electricity supply is 0.8 and the mean time of electricity interruption is 5 hours. Calculate the mean number of annual down hours (i.e., service outage hours) per customer.

In this case, mean time to failure (i.e., the average time between electricity interruptions) is

$$MTTF = \frac{1}{0.8} = 1.25 \text{ hours}$$

By substituting the calculated value and the specified values into Equation (8.6), we get

$$\theta_6 = (8760)(5)/[(1.25)(8760)+(5)]$$

$$= 3.998 \text{ hours per year per customer}$$

Thus, the mean number of annual down hours (i.e., service outage hours) per customer is 3.998 hours.

8.4 LOSS OF LOAD PROBABILITY (LOLP)

Over the years, LOLP has been used as the single most important metric to estimate overall power system reliability and it may simply be described as a projected value of how much time, in the long run, the load on a given power system is expected to be greater than the capacity of the generating resources [8]. Various probabilistic methods/techniques are used for calculating LOLP.

In the setting up of an LOLP criterion, it is always assumed that an electric power system strong enough to have a low LOLP, can probably withstand most of the fore-seeable peak loads, contingencies, and outages. Thus, a utility is always expected to arrange for resources (i.e., generation, load management, purchases, etc.) in such a way so that the resulting system LOLP will be at or lower than an acceptable level.

Generally, the common practice is to plan to power system for achieving an LOLP of 0.1 days per year or lower. All in all, past experiences over the years indicate that there are many difficulties with this use of LOLP. Some of these are as follows [8]:

- Major loss-of-load incidents usually occur because of contingencies not modeled effectively by the traditional LOLP calculation.
- Different LOLP estimation techniques can result in different indices for exactly the same electric power system.
- LOLP does not take into consideration the factor of additional emergency support that one region or control area may receive from another, or other emergency measures/actions that control area operators can exercise for maintaining system reliability.
- LOLP itself does not state the magnitude or duration of the electricity's shortage.

8.5   AVAILABILITY ANALYSIS OF A SINGLE POWER GENERATOR UNIT

There are a number of mathematical models that can be used for performing availability analysis of a single generator unit. Three of these mathematical models are presented below.

8.5.1   MODEL I

This mathematical model represents a single power generator unit that can either be in operating state or failed state. The failed power generator unit is repaired. The power generator unit state-space diagram is shown in Figure 8.1. The numerals in the boxes denote the power generator unit state.

The model is subjected to the following assumptions:

- The power generator unit failure and repair rates are constant.
- The repaired power generator unit is as good as new.
- The power generator unit failures are statistically independent.

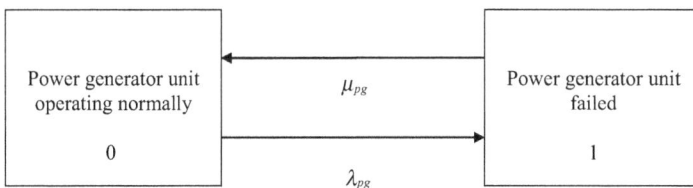

FIGURE 8.1   Power generator unit state-space diagram.

The following symbols are associated with Figure 8.1 diagram and its associated equations:

$P_j(t)$ is the probability that the power generator unit is in state j at time; for $j = 0$ (operating normally), $j = 1$ (failed).
λ_{pg} is the power generator unit failure rate.
μ_{pg} is the power generator unit repair rate.

Using the Markov method presented in Chapter 4, we write down the following equations for Figure 8.1 state-space diagram [11]:

$$\frac{dP_0(t)}{dt} + \lambda_{pg}P_0(t) - \mu_{pg}P_1(t) = 0 \tag{8.7}$$

$$\frac{dP_1(t)}{dt} + \mu_{pg}P_1(t) - \lambda_{pg}P_0(t) = 0 \tag{8.8}$$

At time $t = 0$, $P_0(0) = 1$, and $P_1(0) = 0$.

Solving Equations (8.7)–(8.8) by using Laplace transforms we obtain

$$P_0(t) = \frac{\mu_{pg}}{\lambda_{pg} + \mu_{pg}} + \frac{\lambda_{pg}}{\lambda_{pg} + \mu_{pg}} e^{-\left(\lambda_{pg} + \mu_{pg}\right)t} \tag{8.9}$$

$$P_1(t) = \frac{\lambda_{pg}}{\lambda_{pg} + \mu_{pg}} - \frac{\mu_{pg}}{\lambda_{pg} + \mu_{pg}} e^{-\left(\lambda_{pg} + \mu_{pg}\right)t} \tag{8.10}$$

The power generator unit availability and unavailability are given by

$$AU_{pg}(t) = P_0(t) = \frac{\mu_{pg}}{\lambda_{pg} + \mu_{pg}} + \frac{\mu_{pg}}{\lambda_{pg} + \mu_{pg}} e^{-\left(\lambda_{pg} + \mu_{pg}\right)t} \tag{8.11}$$

and

$$UA_{pg}(t) = P_1(t) = \frac{\lambda_{pg}}{\lambda_{pg} + \mu_{pg}} - \frac{\mu_{pg}}{\lambda_{pg} + \mu_{pg}} e^{-\left(\lambda_{pg} + \mu_{pg}\right)t} \tag{8.12}$$

where

$AV_{pg}(t)$ is the power generator unit availability at time t.
$UA_{pg}(t)$ is the power generator unit unavailability at time t.

For large t, Equations (8.11)–(8.12) reduce to

$$AV_{pg} = \frac{\mu_{pg}}{\lambda_{pg} + \mu_{pg}} \tag{8.13}$$

and

$$UA_{pg} = \frac{\lambda_{pg}}{\lambda_{pg} + \mu_{pg}} \tag{8.14}$$

where

AV_{pg} is the power generator unit steady state availability.
UA_{pg} is the power generator unit steady state unavailability.

Since $\lambda_{pg} = \dfrac{1}{MTTF_{pg}}$ and $\mu_{pg} = \dfrac{1}{MTTR_{pg}}$, Equations (8.13)–(8.14) become

$$AV_{pg} = \frac{MTTF_{pg}}{MTTR_{pg} + MTTF_{pg}} \tag{8.15}$$

and

$$UA_{pg} = \frac{MTTR_{pg}}{MTTR_{pg} + MTTF_{pg}} \tag{8.16}$$

where

$MTTF_{pg}$ is the power generator unit mean time to failure.
$MTTR_{pg}$ is the power generator unit mean time to repair.

Example 8.2

Assume that constant failure and repair rates of a power generator unit are $\lambda_{pg} = 0.0005$ failures/hour and $\mu_{pg} = 0.0008$ repairs/hour, respectively. Calculate the steady state unavailability of the power generator unit.

By substituting the given data values into Equation (8.14), we obtain

$$UA_{pg} = \frac{0.0005}{0.0005 + 0.0008} = 0.3846$$

Thus, the steady state unavailability of the power generator unit is 0.3846.

8.5.2 MODEL II

This mathematical model represents a power generator unit that can be either operating normally (i.e., generating electricity at its full capacity), derated (i.e., generating electricity at a derated capacity, e.g., say 300 megawatts instead of 500 megawatts at full capacity), or failed. This is depicted by the state-space diagram shown in Figure 8.2. The numerals in the boxes denote system state.

The model is subjected to the following assumptions:

- All power generator unit failure and repair rates are constant.
- The power generator unit failures are statistically independent.
- The repaired power generator unit is as good as new.

The following symbols are associated with Figure 8.2 diagram and its associated equations:

$P_j(t)$ is the probability that the power generator unit is in state j at time t; for $j = 0$ (operating normally), $j = 1$ (derated), $j = 2$ (failed).
λ_p is the power generator unit failure rate from state 0 to state 2.
λ_{pd} is the power generator unit failure rate from state 0 to state 1.
λ_{p1} is the power generator unit failure rate from state 1 to state 2.
μ_p is the power generator unit repair rate from state 2 to state 0.
μ_{pd} is the power generator unit repair rate from state 1 to state 0.
μ_{p1} is the power generator unit repair rate from state 2 to state 4.

Using the Markov method presented in Chapter 4, we write down the following equations for Figure 8.2 state-space diagram [11]:

$$\frac{dP_0(t)}{dt} + \left(\lambda_{pd} + \lambda_p\right)P_0(t) - \mu_{pd}P_1(t) - \mu_p P_2(t) = 0 \qquad (8.17)$$

$$\frac{dP_1(t)}{dt} + \left(\mu_{pd} + \lambda_{p1}\right)P_1(t) - \mu_{p1}P_2(t) - \lambda_{pd}P_0(t) = 0 \qquad (8.18)$$

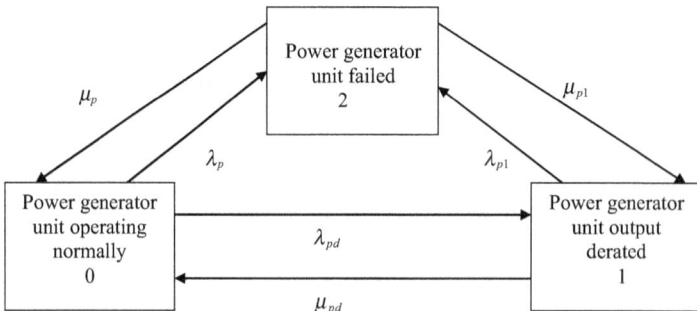

FIGURE 8.2 Power generator unit state-space diagram.

$$\frac{dP_2(t)}{dt} + \left(\mu_p + \mu_{p1}\right)P_2(t) - \lambda_{p1}P_1(t) - \lambda_p P_0(t) = 0 \qquad (8.19)$$

At time $t = 0$, $P_0(0) = 1$, $P_1(0) = 0$, and $P_2(0) = 0$.

Solving Equations (8.17)–(8.19) by using Laplace transforms, we obtain

$$P_0(t) = \frac{B_1}{c_1 c_2} + \frac{B_2}{c_1(c_1 - c_2)}e^{c_1 t} + \left[1 - \frac{B_1}{c_1 c_2} - \frac{B_2}{c_1(c_1 - c_2)}\right]e^{c_2 t} \qquad (8.20)$$

where

$$B_1 = \mu_p \mu_{pd} + \lambda_{p1}\mu_p + \mu_{pd}\mu_{p1} \qquad (8.21)$$

$$B_2 = \mu_{pd}c_1 + \mu_p c_1 + \mu_{p1}c_1 + c_1\lambda_{p1} + c_1^2 + \mu_{pd}\mu_p + \lambda_{p1}\mu_p + \mu_{pd}\mu_{p1} \qquad (8.22)$$

$$c_1, c_2 = \frac{-B_3\left[B_3^2 - 4B_4\right]^{1/2}}{2} \qquad (8.23)$$

$$B_3 = \mu_p + \mu_{p1} + \mu_{pd} + \lambda_p + \lambda_{p1} + \lambda_{pd} \qquad (8.24)$$

$$B_4 = \mu_{pd}\mu_p + \lambda_{p1}\mu_p + \mu_{pd}\mu_{p1} + \mu_p\lambda_{pd} + \lambda_{p1}\lambda_{pd} + \mu_{pd}\lambda_p$$
$$+ \lambda_p\mu_{p1} + \lambda_p\lambda_{p1} + \lambda_{pd}\mu_{p1} \qquad (8.25)$$

$$P_1(t) = \frac{B_5}{c_1 c_2} + \frac{B_6}{c_1(c_1 - c_2)}e^{c_1 t} - \left[\frac{B_5}{c_1 c_2} + \frac{B_6}{c_1(c_1 - c_2)}\right]e^{c_2 t} \qquad (8.26)$$

where

$$B_5 = \lambda_{pd}\mu_p + \lambda_{pd}\mu_{p1} + \lambda_p\mu_{p1} \qquad (8.27)$$

$$B_6 = c_1\lambda_{pd} + B_5 \qquad (8.28)$$

$$P_2(t) = \frac{B_7}{c_1 c_2} + \frac{B_8}{c_1(c_1 - c_2)}e^{c_1 t} - \left[\frac{B_7}{c_1 c_2} + \frac{B_8}{c_1(c_1 - c_2)}\right]e^{c_2 t} \qquad (8.29)$$

where

$$B_7 = \lambda_{pd}\lambda_{p1} + \mu_{pd}\lambda_p + \lambda_p\lambda_{p1} \tag{8.30}$$

$$B_8 = c_1\lambda_p + B_7 \tag{8.31}$$

The power generator unit operational availability is expressed by

$$AV_{pg}(t) = P_0(t) + P_1(t) \tag{8.32}$$

For large t, Equation (8.32) reduces to

$$AV_{pg}(t) = \lim_{t\to\infty}\left[P_0(t) + P_1(t)\right] = \frac{B_1 + B_6}{c_1 c_2} \tag{8.33}$$

where

AV_{pg} is the power generator unit operational steady state availability.

8.5.3 Model III

This mathematical model represents a power generator unit that can either be in operating state or failed state or down for preventive maintenance. This is depicted by the state-space diagram shown in Figure 8.3. The numerals in the boxes and circle denote the system state.

The model is subjected to the following assumptions:

- The power generator unit failure, repair, preventive maintenance down, and preventive maintenance performance rates are constant.
- After preventive maintenance and repair, the power generator unit is as good as new.
- The power generator unit failures are statistically independent.

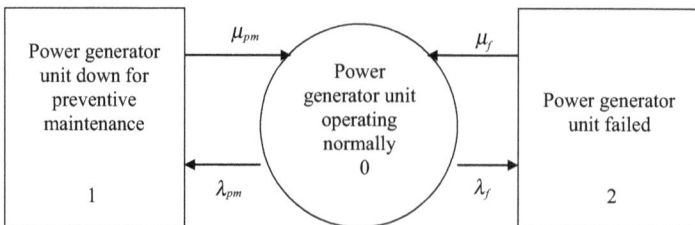

FIGURE 8.3   Power generator unit state-space diagram.

The following symbols are associated with the state-space diagram shown in Figure 8.3 and its associated equations:

$P_j(t)$ is the probability that the power generator unit is in state j at time t; for $j = 0$
  (operating normally), $j = 1$ (down for preventive maintenance), $j = 2$ (failed).
λ_f is the power generator unit failure rate.
λ_{pm} is the power generator unit (down for) preventive maintenance rate.
μ_f is the power generator unit repair rate.
μ_{pm} is the power generator unit preventive maintenance performance (repair) rate.

Using the Markov method presented in Chapter 4, we write down the following equations for Figure 8.3 state-space diagram [11]:

$$\frac{dP_0(t)}{dt} + \left(\lambda_{pm} + \lambda_f \right) P_0(t) - \mu_f P_2(t) - \mu_{pm} P_1(t) = 0 \tag{8.34}$$

$$\frac{dP_1(t)}{dt} + \mu_{pm} P_1(t) - \lambda_{pm} P_0(t) = 0 \tag{8.35}$$

$$\frac{dP_2(t)}{dt} + \mu_f P_2(t) - \lambda_f P_0(t) = 0 \tag{8.36}$$

At time $t = 0$, $P_0(0) = 1, P_1(0) = 0,$ and $P_2(0) = 0.$

Solving Equations (8.34)–(8.36) by using Laplace transforms, we obtain

$$P_0(t) = \frac{\mu_{pm}\mu_f}{k_1 k_2} + \left[\frac{\left(k_1 + \mu_{pm} \right) \left(k_1 + \mu_f \right)}{k_1 \left(k_1 - k_2 \right)} \right] e^{k_1 t} - \left[\frac{\left(k_2 + \mu_{pm} \right) \left(k_2 + \mu_f \right)}{k_2 \left(k_1 - k_2 \right)} \right] e^{k_2 t} \tag{8.37}$$

$$P_1(t) = \frac{\lambda_{pm}\mu_f}{k_1 k_2} + \left[\frac{\left(\lambda_{pm} k_1 + \lambda_{pm} \right)}{k_1 \left(k_1 - k_2 \right)} \right] e^{k_1 t} - \left[\frac{\left(\mu_f + k_2 \right) \lambda_{pm}}{k_2 \left(k_1 - k_2 \right)} \right] e^{k_2 t} \tag{8.38}$$

$$P_2(t) = \frac{\lambda_f \mu_{pm}}{k_1 k_2} + \left[\frac{\left(\lambda_f k_1 + \lambda_f \mu_{pm} \right)}{k_1 \left(k_1 - k_2 \right)} \right] e^{k_1 t} - \left[\frac{\left(\mu_{pm} + k_2 \right) \lambda_f}{k_2 \left(k_1 - k_2 \right)} \right] e^{k_2 t} \tag{8.39}$$

where

$$k_1 k_2 = \mu_{pm}\mu_f + \lambda_{pm}\mu_f + \lambda_f \mu_{pm} \tag{8.40}$$

$$k_1 + k_2 = -\left(\mu_{pm} + \mu_f + \lambda_{pm} + \lambda_s \right) \tag{8.41}$$

The power generator unit availability, $AV_{pg}(t)$, is given by

$$AV_{pg}(t) = P_0(t) = \frac{\mu_{pm}\mu_f}{k_1 k_2} + \left[\frac{(k_1 + \mu_{pm})(k_1 + \mu_f)}{k_1(k_1 - k_2)} \right] e^{k_1 t} - \left[\frac{(k_2 + \mu_{pm})(k_1 + \mu_f)}{k_2(k_1 - k_2)} \right] e^{k_2 t}$$

(8.42)

It is to be noted that the above availability expression is valid if and only if k_1 and k_2 are negative. Thus, for large t, Equation (8.42) reduces to

$$AV_{pg} = \lim_{t \to \infty} AV_{pg}(t) = \frac{\mu_{pm}\mu_f}{k_1 k_2}$$

(8.43)

where

AV_{pg} is the power generator unit steady state availability.

Example 8.3

Assume that for a power generator unit we have the following data values:

$\lambda_f = 0.0001$ failures/hour
$\mu_f = 0.0004$ repairs/hour
$\lambda_{pm} = 0.0006$/hour
$\mu_{pm} = 0.0008$/hour

Calculate the power generator unit steady state availability.
By inserting the stated data values into Equation (8.43), we obtain

$$AV_{pg} = \frac{(0.0008)(0.0004)}{(0.0008)(0.0004) + (0.0006)(0.0004) + (0.0001)(0.0008)}$$
$$= 0.5$$

Thus, the steady state availability of the power generator unit is 0.5.

8.6 AVAILABILITY ANALYSIS OF TRANSMISSION AND ASSOCIATED SYSTEMS

In the area of power system, various types of systems and equipment are used for transmitting electrical energy from one point to another. Two examples of such systems and equipment are transmission lines and transformers. This section presents two mathematical models to perform availability analysis of such systems [4,7,9,11].

8.6.1 MODEL I

This mathematical model represents a system composed of two non-identical and redundant transmission lines subject to the occurrence of common-cause failures. A common-cause failure may simply be described as any instance where multiple units fail due to a single cause [7,13,14]. In transmission lines, a common cause failure may take place due to factors such as poor weather, aircraft crash, and tornado. The system state-space diagram is shown in Figure 8.4.

The numerals in the circles and boxes denote system states. The model is subjected to the following assumptions:

- All failure and repair rates of transmission lines are constant.
- A repaired transmission line is as good as new.
- All failures are statistically independent.

The following symbols are associated with the state-space diagram shown in Figure 8.4 and its associated equations:

$P_j(t)$ is the probability that the system is in state j at time t; for $j = 0$ (both transmission lines operating normally), $j = 1$ (transmission line 1 failed, other operating), $j = 2$ (transmission line 2 failed, other operating), $j = 3$ (both transmission lines failed).

λ_{t1} is the transmission line 1 failure rate.

λ_{t2} is the transmission line 2 failure rate.

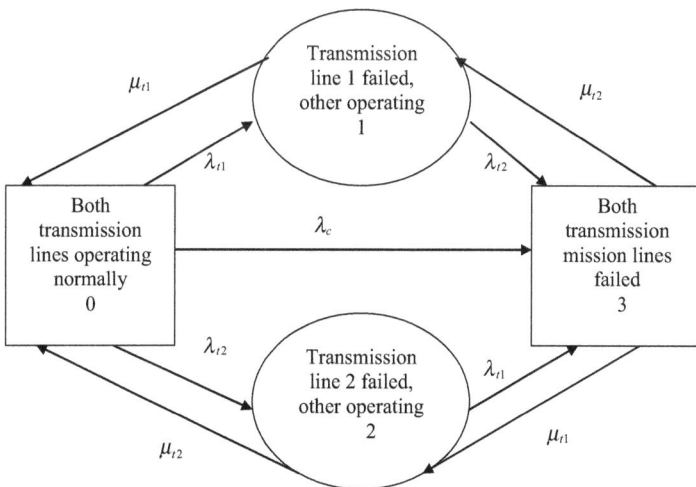

FIGURE 8.4   State-space diagram for two non-identical and redundant transmission lines.

λ_c is the system common-cause failure rate.
μ_{r1} is the transmission line 1 repair rate.
μ_{r2} is the transmission line 2 repair rate.

Using the Markov method presented in Chapter 4, we write down the following equations for Figure 8.4 state-space diagram [11,13,15]:

$$\frac{dP_0(t)}{dt}+\left(\lambda_{t1}+\lambda_{t2}+\lambda_c\right)P_0(t)-\mu_{r1}P_1(t)-\mu_{r2}P_2(t)=0 \tag{8.44}$$

$$\frac{dP_1(t)}{dt}+\left(\lambda_{t2}+\mu_{r1}\right)P_1(t)-\mu_{r2}P_3(t)-\lambda_{t1}P_0(t)=0 \tag{8.45}$$

$$\frac{dP_2(t)}{dt}+\left(\lambda_{t1}+\mu_{r2}\right)P_2(t)-\mu_{r1}P_3(t)-\lambda_{t2}P_0(t)=0 \tag{8.46}$$

$$\frac{dP_3(t)}{dt}+\left(\mu_{r1}+\mu_{r2}\right)P_3(t)-\lambda_{t1}P_2(t)-\lambda_{t2}P_1(t)-\lambda_c P_0(t)=0 \tag{8.47}$$

At time $t=0$, $P_0(0)=1, P_1(0)=0, P_2(0)=0,$ and $P_3(0)=0.$

The following steady-state equations are obtained from Equations (8.44)–(8.47) by setting the derivatives with respect to time t equal to zero and using the relationship $\sum_{j=0}^{3}P_j=1$:

$$P_0=\mu_{r1}\mu_{r2}D/D_3 \tag{8.48}$$

where

$$D=D_1+D_2 \tag{8.49}$$

$$D_1=\left(\lambda_{t1}+\mu_{r1}\right) \tag{8.50}$$

$$D_2=\left(\lambda_{t2}+\mu_{r2}\right) \tag{8.51}$$

$$D_3=DD_1D_2+\lambda_c\left[D_1\left(D_2+\mu_{r1}\right)+\mu_{r2}D_2\right] \tag{8.52}$$

$$P_1=\left[D_1\lambda_{t1}+D_4\lambda_c\right]\mu_{t2}/D_3 \tag{8.53}$$

where

$$D_4=\left(\lambda_{t1}+\mu_{t2}\right) \tag{8.54}$$

$$P_2 = \left[D\lambda_{t2} + D_5\lambda_c \right]\mu_{t1} / D_3 \qquad (8.55)$$

where

$$D_5 = \left(\lambda_{t2} + \mu_{t1} \right) \qquad (8.56)$$

$$P_3 = D\lambda_{t1}\lambda_{t2} + D_4 D_5 \lambda_c / D_3 \qquad (8.57)$$

$P_0, P_1, P_2, and\ P_3$ are the steady state probabilities of the system being in states 0, 1, 2, and 3, respectively.

The system steady state availability and unavailability are given by

$$AV_{ss} = P_0 + P_1 + P_2 \qquad (8.58)$$

and

$$UAV_{ss} = P_3 \qquad (8.59)$$

where

AV_{ss} is the system steady state availability.
UAV_{ss} is the system steady state unavailability.

8.6.2 Model II

This mathematical model represents a system composed of transmission lines and other equipment operating in fluctuating outdoor environments (i.e., normal and stormy). The system can malfunction under both these conditions. The system state-space diagram is shown in Figure 8.5. The numerals in circles and boxes denote system states.
The model is subjected to the following assumptions:

- All failure, repair, and weather fluctuation rates are constant.
- The repaired system is as good as new.
- All failures are statistically independent.

The following symbols are associated with the state-space diagram shown in Figure 8.4 and its associated equations:

$P_j(t)$ is the probability that the system is in state j at time t; for $j = 0$ (operating normally in normal weather), $j = 1$ (failed in normal weather), $j = 2$ (operating normally in stormy weather), $j = 3$ (failed in stormy weather).
λ_{nw} is the system constant failure rate in normal weather.
λ_{sw} is the system constant failure rate in stormy weather.
μ_{nw} is the system constant repair rate in normal weather.

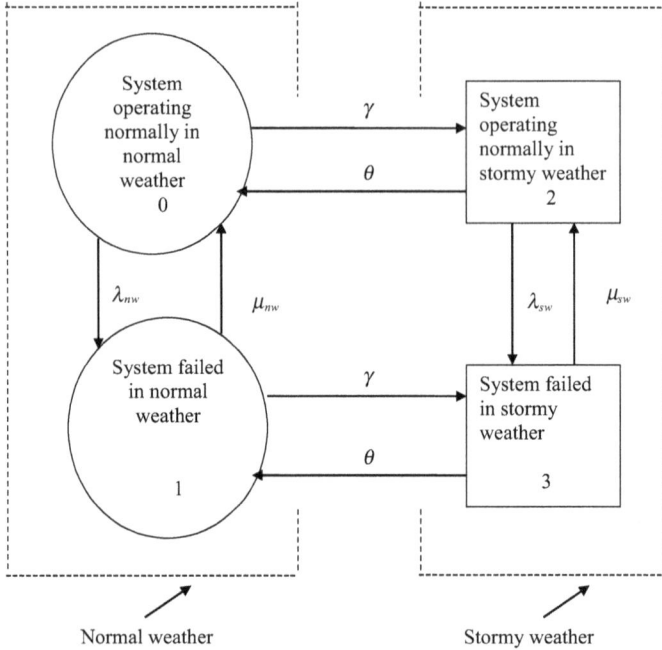

FIGURE 8.5   State-space diagram of a system operating in fluctuating outdoor environments.

μ_{sw} is the system constant repair rate in stormy weather.

γ is the constant transition rate from normal weather to stormy weather.

θ is the constant transition rate from stormy weather to normal weather.

Using the Markov method presented in Chapter 4, we write down the following equations for Figure 8.4 state-space diagram [11,15]:

$$\frac{P_0(t)}{dt} + (\lambda_{nw} + \gamma) P_0(t) - \theta P_2(t) - \mu_{nw} P_1(t) = 0 \tag{8.60}$$

$$\frac{dP_1(t)}{dt} + (\mu_{nw} + \gamma) P_1(t) - \theta P_3(t) - \lambda_{nw} P_0(t) = 0 \tag{8.61}$$

$$\frac{dP_2(t)}{dt} + (\lambda_{sw} + \theta) P_2(t) - \mu_{sw} P_3(t) - \gamma P_0(t) = 0 \tag{8.62}$$

$$\frac{dP_3(t)}{dt} + (\theta + \mu_{sw}) P_3(t) - \lambda_{sw} P_2(t) - \gamma P_1(t) = 0 \tag{8.63}$$

At time $t = 0$, $P_0(0) = 1$, $P_1(0) = 0$, $P_2(0) = 0$, $and\ P_3(0) = 0$.

The following steady-state equations are obtained from Equations (8.60)–(8.63) by setting the derivatives with respect to time t equal to zero and using the relationship $\sum_{j=0}^{3} P_j = 1$:

$$P_0 = \frac{\theta A_1}{\gamma\left(A_2 + A_3\right) + \theta\left(A_4 + A_1\right)} \tag{8.64}$$

where

$$A_1 = \mu_{sw}\gamma + \mu_{nw}A_5 \tag{8.65}$$

$$A_2 = \mu_{nw}\theta + \mu_s A_6 \tag{8.66}$$

$$A_3 = \lambda_{nw}\theta + \lambda_{sw}A_6 \tag{8.67}$$

$$A_4 = \lambda_{sw}\gamma + \lambda_{nw}A_5 \tag{8.68}$$

$$A_5 = \lambda_{sw} + \theta + \mu_{sw} \tag{8.69}$$

$$A_6 = \lambda_{nw} + \gamma + \mu_{nw} \tag{8.70}$$

$$P_1 = A_4 P_0 / A_1 \tag{8.71}$$

$$P_2 = \gamma P_0 A_2 / \theta A_1 \tag{8.72}$$

$$P_3 = \gamma P_0 A_3 / \theta A_1 \tag{8.73}$$

$P_0, P_1, P_2,$ and P_3 are the steady state probabilities of the system being in states 0, 1, 2, and 3, respectively.

The system steady state availability and unavailability are given by

$$AV_{ss} = P_0 + P_2 \tag{8.74}$$

and

$$UAV_{ss} = P_1 + P_3 \tag{8.75}$$

where

AV_{ss} and UAV_{ss} are the system steady state availability and unavailability, respectively.

8.7 PROBLEMS

1. Define the following terms:
   - Forced outage
   - Forced outage rate
   - Power system reliability
2. Write an essay on power system reliability.
3. Define the following indices:
   - Average service availability index
   - Customer average interruption duration index
   - System average interruption frequency index
4. Describe loss of load probability (LOLP).
5. What are the difficulties associated with the use of LOLP?
6. Assume that the annual failure rate of the electricity supply is 0.9 and the mean time to electricity interruption is 4 hours. Calculate the mean number of annual down hours (i.e., service outage hours) per customer.
7. Assume that constant failure and repair rates of a power generator unit are 0.0004 failures/hour and 0.0006 repairs/hour, respectively. Calculate the steady state availability of the power generator unit.
8. Assume that for a power generator unit we have the following data values:
   - $\lambda_f = 0.0002$ failures/hour
   - $\lambda_{pm} = 0.0008 /$ hour
   - $\mu_f = 0.0006$ repairs / hour
   - $\mu_{pm} = 0.0009 /$ hour

   Calculate the power generator unit steady state unavailability by using Equation (8.43).
9. Prove that the sum of Equations (8.48), (8.53), (8.55), and (8.57) is equal to unity.
10. Prove Equations (8.48), (8.53), (8.55), and (8.57).

REFERENCES

1. Billinton, R., Allan, R.N., Reliability of Electric Power Systems: An Overview, in *Handbook of Reliability Engineering*, edited by H. Pham, Springer-Verlag, London, 2003, pp. 511–528.
2. Layman, W.J., Fundamental consideration in preparing a master system plan, *Electrical World*, Vol. 101, 1933, pp. 778–792.
3. Smith, S.A., Service reliability measured by probabilities of outage, *Electrical World*, Vol. 103, 1934, pp. 371–374.
4. Billinton, R., *Power System Reliability Evaluation*, Gordon and Breach Science Publishers, New York, 1970.
5. Dhillon, B.S., *Power System Reliability, Safety, and Management*, Ann Arbor Science, Ann Arbor, Michigan, 1983.
6. Bilinton, R., Bibliography on the application of probability methods in power system reliability evaluation, *IEEE Transactions on Power Apparatus and Systems*, Vol. 91, 1972, pp. 649–660.

7. Dhillon, B.S., *Applied Reliability and Quality: Fundamentals, Methods, and Procedures*, Springer-Verlag, London, 2007.
8. Kueck, J.D., Kirby, B.J., Overholt, P.N., Markel, L.C., *Measurement Practices for Reliability and Power Quality, Report No. ORNL/TM-2004/91*, June 2004. Available from the Oak Ridge National Laboratory, Oak Ridge, Tennessee, USA.
9. Endrenyi, J., *Reliability Modeling in Electric Power Systems*, John Wiley, New York, 1978.
10. Kennedy, B., *Power Quality Primer*, McGraw Hill, New York, 2000.
11. Dhillon, B.S., *Reliability Engineering in Systems Design and Operation*, Van Nostrand Reinhold, New York, 1983.
12. Gangel, M.W., Ringlee, R.J., Distribution system reliability performance, *IEEE Transactions on Power Apparatus and Systems*, Vol. 87, 1968, pp. 1657–1665.
13. Billinton, R., Medicherala, T.L.P., Sachdev, M.S., Common-cause outages in multiple circuit transmission lines, *IEEE Transactions on Reliability*, Vol. 27, 1978, pp. 128–131.
14. Gangloff, W.C., Common mode failure analysis, *IEEE Transactions on Power Apparatus and Systems*, Vol. 94, Feb. 1975, pp. 27–30.
15. Dhillon, B.S., Singh, C., *Engineering Reliability: New Techniques and Applications*, John Wiley, New York, 1981.

9 Medical Equipment Reliability

9.1 INTRODUCTION

The earliest use of medical devices' history may be traced back to ancient Etruscans and Egyptians using various types of dental devices [1]. Nowadays, medical devices and equipment are widely used around the globe. In fact, in 1988, the production of medical equipment around the globe was estimated to be around $36 billion [1] and in 1997, the global market for medical devices was valued at approximately $120 billion [2].

In modern times, the beginning of the medical equipment or device reliability field may be traced back to the latter part of the 1960s, when a number of publications on the subject appeared [3–7]. These publications covered topics such as *Safety and Reliability in Medical Electronics*, *Reliability of ECG Instrumentation*, and *Some Instrument Induced Errors in the Electrocardiogram* [3–5]. In 1980, an article presented a comprehensive list of medical equipment reliability-related publications [8] and in 2000, a book entitled *Medical Device Reliability and Associated Areas* provided a comprehensive list of publications on the subject [9].

This chapter presents various important aspects of medical equipment reliability.

9.2 MEDICAL EQUIPMENT RELIABILITY-RELATED FACTS AND FIGURES

There are many, directly or indirectly, medical equipment reliability-related facts and figures. Some of these are as follows:

- In 1997, there were a total of 10,420 registered medical device manufacturers in the entire United States [10].
- In 1969, the United States Department of Health, Education, and Welfare special committee reported that over a 10 year period, approximately 10,000 injuries were related to medical equipment/devices and 731 resulted in fatalities [11,12].
- The Emergency Care Research Institute (ECRI) tested a sample of 15,000 products used in hospitals and discovered that approximately 4% to 6% of these products were sufficiently dangerous to warrant immediate correction [13].

- In 1990, a study conducted by the US Food and Drug Administration (FDA) revealed that approximately 44% of the quality-associated problems that resulted in the voluntary recall of medical devices for the period October 1983 to September 1989, were due to errors/deficiencies that could have been prevented through effective design controls [14].
- Due to faulty medical instrumentation, approximately 1,200 deaths per year occur in the United States [15,16].
- A study reported that approximately 100,000 Americans die each year due to human errors, and their financial impact on the United States economy was estimated to be somewhere between $17 billion and $29 billion [17].
- A study reported that over 50% of all technical-related medical equipment problems were directly or indirectly due to operator errors [13].

9.3 MEDICAL DEVICES AND MEDICAL EQUIPMENT/DEVICES CATEGORIES

Nowadays, there are over 5,000 different types of medical devices being used in a modern hospital and they range from a simple tongue depressor to a complex pacemaker [1,9]. Thus, the criticality of these devices' reliability varies from one device to another. Nonetheless, past experiences over the years clearly indicate that the failure of medical devices has been very costly in terms of dollars and cents, fatalities, injuries, etc. Needless to say, modern medical devices and equipment have become highly sophisticated and complex and are expected to function under stringent environments.

Electronic equipment/devices used in the health care system may be categorized under the following three classifications [3]:

- **Classification I.** This classification includes those medical equipment/devices that are immediately and directly responsible for the patient's life or may become so under emergency situations. When such equipment/devices malfunction, there is seldom sufficient time for the repair action. Thus, this type of equipment/devices must always function successfully at the moment of need. Four examples of such equipment/devices are as follows:
  - Cardiac defibrillators
  - Cardiac pacemakers
  - Respirators
  - Electrocardiographic monitors.
- **Classification II.** This classification includes those medical devices/equipment that are not critical to a patient's welfare or life but just serve as convenience devices/equipment. Some examples of such devices/equipment are as follows:
  - Wheelchairs
  - Bedside television sets
  - Electric beds.
- **Classification III.** This classification includes those medical devices/ equipment that are utilized for routine or semi-emergency diagnostic or therapeutic purposes. Failure of such devices/equipment is not as critical as those

fall under Classification I, because there is time for repair. Six examples of such devices/equipment are as follows:

- Colorimeters
- Spectrophotometers
- Diathermy equipment
- Gas analyzers
- Ultrasound equipment
- Electrocardiograph and electroencephalograph recorders and monitors

Finally, it is to be noted that there could be some overlap between the above three classifications of devices/equipment, particularly between classifications I and III. A typical example of such equipment/devices is an electrocardiograph recorder or monitor.

9.4 METHODS AND PROCEDURES FOR IMPROVING RELIABILITY OF MEDICAL EQUIPMENT

There are many methods and procedures used for improving medical equipment reliability. Some of these are presented below.

9.4.1 PARTS COUNT METHOD

This method is used for predicting equipment/system failure during the bid proposal and early design stages [18]. The method requires information on three areas shown in Figure 9.1.

The parts count method calculates the equipment/system failure rate under the single-use environment by using the following equation [18]:

$$\lambda_e = \sum_{j=1}^{k} \theta_j \left(\lambda_{gp} Q_{gp} \right)_j \tag{9.1}$$

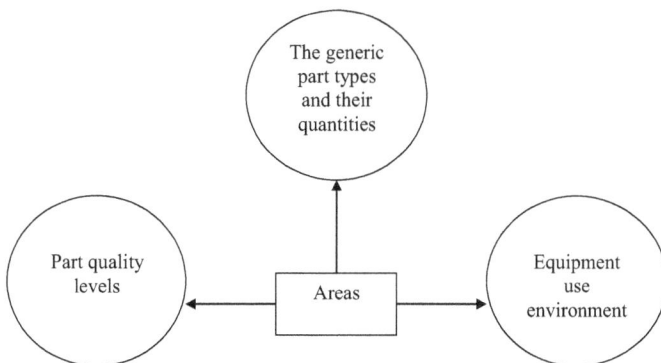

FIGURE 9.1 Areas of information required by the parts count method.

where

λ_e is the equipment/system failure rate expressed in failures/10^6 hours.
k is the number of different generic part/component classifications.
θ_j is the generic part quantity for classification j.
λ_{gp} is the generic part failure rate expressed in failures/10^6 hours.
Q_{gp} is the generic component quality factor.

The values of λ_{gp} and Q_{gp} are tabulated in Ref. [18], and additional information on this method is available in Refs. [18,19].

9.4.2 FAILURE MODE AND EFFECT ANALYSIS (FMEA)

This method is widely used for evaluating design at the early stage from the reliability aspect. This criterion is extremely useful for identifying the need for and the effects of design change. The method requires the listing of all possible failure modes of each and every component/part on paper and their effect on the listed subsystems, etc. FMEA is known as failure modes, effects, and criticality analysis (FMECA) when criticalities or priorities are assigned to failure mode effects.

Some of the FMEA's important characteristics are as follows [20]:

- It is an effective tool for identifying weak spots in system design and indicate areas where further or detailed analysis is required.
- By examining failure effects of all components/parts, the entire system is screened completely.
- It is an upward approach that begins at the detailed level.

Additional information on this method is available in Chapter 4 and in Refs. [20,21].

9.4.3 GENERAL APPROACH

This is a 13-step approach developed by Bio-Optronics for producing reliable and safe medical devices [22]. The approach steps are as follows [22]:

- **Step 1:** Conduct analysis of existing medical problems.
- **Step 2:** Develop a product concept for determining a solution to a specific medical-associated problem.
- **Step 3:** Evaluate all possible environments under which the medical device under consideration is functioning.
- **Step 4:** Evaluate all possible individuals expected to operate the product/device under consideration.
- **Step 5:** Construct a prototype.
- **Step 6:** Test the prototype under laboratory environment.
- **Step 7:** Test the prototype under the actual field use environment.
- **Step 8:** Make appropriate changes to the product/device design to satisfy field requirements.

- **Step 9:** Conduct Laboratory and field test on the modified version of the product/device.
- **Step 10:** Build pilot units for performing necessary tests.
- **Step 11:** Ask impartial experts for testing pilot units under the field use environments.
- **Step 12:** Release the product/device design for production.
- **Step 13:** Study the product/device field performance and support with appropriate product/device maintenance.

9.4.4 FAULT TREE ANALYSIS (FTA)

FTA starts by identifying an undesirable event, called the top event, associated with a system under consideration [23]. Fault events which could cause the top event's occurrence are generated and connected by logic operators such as OR and AND. The OR gate provides a TRUE (failure) output when only one or more of its inputs are true (failures). In contrast, the AND gate provides a TRUE (failed) output when all its inputs are TRUE (failures). All in all, the construction of the fault tree proceeds by generation of fault events in a successive manner until the fault events need not be developed any further.

Additional information on FTA is available in Chapter 4 and in Refs. [23,24].

9.4.5 MARKOV METHOD

This method is a very general approach and it can generally handle more cases than any other technique or method. It can be utilized in situations when the parts/components are independent as well as for systems/equipment involving dependent failure and repair modes.

This method proceeds by the enumeration of system states. The state probabilities are calculated, and the steady-state reliability measures can be computed by utilizing the frequency balancing method [25]. Additional information on the method is available in Chapter 4 and in Refs. [26,27].

9.5 HUMAN ERROR IN MEDICAL EQUIPMENT

Human errors are universal and are committed each day around the globe. Past experiences over the years clearly indicate that although most are trivial, some can be quite serious or fatal. In the area of health care, one study reported that in a typical year approximately 100,000 Americans die due to human errors [17]. Nonetheless, some of the medical device/equipment-associated, directly or indirectly, human error facts and figures are as follows:

- Over 50% of all technical medical equipment problems are, directly or indirectly, due to operator errors [13].
- A fatal radiation overdose accident involving the Therac radiation therapy device was the result of a human error [28].

- The Center for Devices and Radiological Health (CDRH) of the Food and Drug Administration reported that human errors account for around 60% of all medical device-associated deaths or injuries in the United States [29].
- A patient was seriously injured by over-infusion because the attending nurse wrongly read the number 7 as 1 [30].
- Human error, directly or indirectly, is responsible for up to 90% of accidents both generally and in medical devices [31,32].

9.5.1 IMPORTANT MEDICAL EQUIPMENT/DEVICE OPERATOR ERRORS

Past experiences over the years indicate that there are many types of operator errors which occur during medical equipment/device operation or maintenance. Some of these are as follows [33]:

- Incorrect selection of devices in regard to the clinical requirements and objectives.
- Mistakes in setting device parameters.
- Departure from following stated procedures and instructions.
- Wrong interpretation of or failure to recognize critical device outputs.
- Wrong decision-making and actions in critical moments.
- Untimely or inadvertent activation of controls.
- Over-reliance on automatic features of equipment/devices.
- Misssembly.

9.5.2 MEDICAL DEVICES WITH HIGH INCIDENCE OF HUMAN ERROR

Over the years, many studies have been conducted to highlight medical devices with a high occurrence of human error. Consequently, the most error-prone medical devices were highlighted. These devices, in the order of least error-prone to most error-prone, are as follows [34]:

- Contact lens cleaning and disinfecting solutions
- Continuous ventilator (respirator)
- External low-energy defibrillator
- Trans-luminal coronary angioplasty catheter
- Catheter guide wire
- Catheter introducer
- Peritoneal dialysate delivery system
- Implantable pacemaker
- Mechanical/hydraulic impotence device
- Non-powered suction apparatus
- Electrosurgical cutting and coagulation device
- Urological catheter
- Infusion pump
- Intra-vascular catheter
- Implantable spinal cord simulator

- Permanent pacemaker electrode
- Administration kit for peritoneal dialysis
- Orthodontic bracket aligner
- Balloon catheter
- Glucose meter

9.6 MEDICAL EQUIPMENT MAINTAINABILITY AND MAINTENANCE

Medical equipment maintainability may simply be described as the probability that a failed piece of medical equipment will be restored to its acceptable operating state. Similarly, medical equipment maintenance is all actions necessary for retaining medical equipment in, or restoring to, a specified condition. Both medical equipment maintainability and maintenance are discussed below, separately [35,36].

9.6.1 MEDICAL EQUIPMENT MAINTAINABILITY

Past experiences over the years clearly indicate that the application of maintainability principles during designing the engineering equipment has helped to produce effectively maintainable end products. Their application in the medical equipment's design can also be quite helpful for producing effectively maintainable end medical items. This section presents three aspects of maintainability considered useful for producing effectively maintainable medical equipment.

9.6.1.1 Reasons for Maintainability Principles' Application

There are many reasons for maintainability principles' application. Some of the main reasons are as follows [37]:

- To determine the number of labor hours and related resources needed for carrying out the projected maintenance
- To lower projected maintenance time
- To determine the amount of downtime due to maintenance
- To lower projected maintenance cost through design modifications.

9.6.1.2 Maintainability Design Factors

There are many maintainability design factors and some of the most frequently addressed factors are shown in Figure 9.2 [38]. Additional information on these factors is available in Refs. [9,38].

9.6.1.3 Maintainability Measures

There are various types of maintainability measures used in conducting maintainability analysis of engineering equipment/system. Two of these measures are as follows [37–39]:

- Mean Time to Repair (*MTTR*)

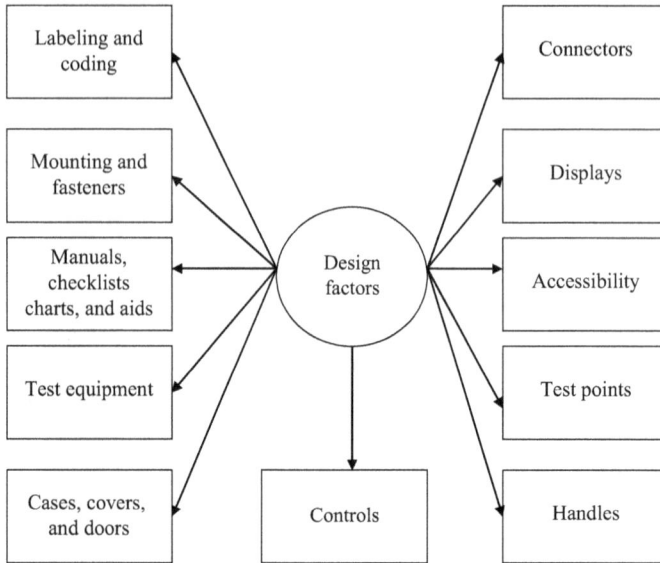

FIGURE 9.2 Frequently addressed maintainability design factors.

It is defined by

$$MTTR = \frac{\sum\limits_{j=1}^{n} t_{rj}\lambda_j}{\sum\limits_{j=1}^{n} \lambda_j}$$

(9.2)

where

n is the number of units.
t_{rj} is the repair time required to repair unit j; for $j = 1,2,3, ..., n$.
λ_j is the constant failure rate of unit j; for $j = 1, 2, 3, ..., n$.

• Maintainability Function

This measure is used for predicting the probability that the repair will be accomplished in a time t, when it starts on an item/equipment at time $t = 0$. Thus, the maintainability function, $m(t)$, is expressed as follows:

$$m(t) = \int_0^t f(t)dt$$

(9.3)

where

> t is the time.
> $f(t)$ is the probability density function of the repair time.

Equation (9.3) is used for obtaining maintainability functions for various probability distributions (e.g., exponential, Weibull, and normal) representing failed item/system/equipment repair times. Maintainability functions for various probability distributions are available in Refs. [38–40].

Example 9.1

Assume that the repair times of a medical equipment/system are exponentially distributed with a mean value (i.e., mean time to repair (MTTR)) of 5 hours. Calculate the probability that a repair will be accomplished in 15 hours.

Thus, in this case, the probability density function of repair times is defined by

$$f(t) = \frac{1}{MTTR} \exp\left(-\frac{t}{MTTR}\right)$$

$$= \frac{1}{5} \exp\left(-\frac{t}{5}\right) \tag{9.4}$$

By inserting Equation (9.4) and the specified data value into Equation (9.3), we obtain

$$M(15) = \int_0^{15} \frac{1}{5} \exp\left(-\frac{t}{5}\right) dt$$

$$= 1 - \exp\left(-\frac{15}{5}\right) = 0.9502$$

Thus, the probability of accomplishing a repair within 15 hours is 0.9502.

9.6.2   MEDICAL EQUIPMENT MAINTENANCE

For the purpose of maintenance and repair, medical equipment may be classified under six classifications [41]:

- **Classification I: Imaging and radiation therapy equipment**. Some examples of such equipment are linear accelerators, X-ray machines, and ultrasound devices.

- **Classification II**: **Patient diagnostic equipment.** Some examples of such equipment are spirometers, endoscopes, and physiologic monitors.
- **Classification III: Patient environmental and transport equipment.** Some examples of such equipment are patient beds, wheelchairs, and patient-room furniture.
- **Classification IV**: **Life support and therapeutic equipment.** Some examples of such equipment are ventilators, lasers, and anesthesia machines.
- **Classification V**: **Laboratory apparatus.** Some examples of such equipment are lab analyzers, lab refrigeration equipment, and centrifuges.
- **Classification VI**: **Miscellaneous equipment.** This classification contains all other items that are not included in the previous five classifications, for example, sterilizers.

9.6.2.1   Indices

Similar to the case of the general maintenance activity, there are many indices that can be used for measuring the effectiveness of the medical equipment maintenance-related activity.

Three of these indices are as follows [41]:

- **Index I**

  This index measures how often the customer has to request for service per medical equipment and is defined by

  $$\alpha_c = \frac{R_{rr}}{n} \tag{9.5}$$

  where

  α_c is the number of repair requests completed per medical equipment.
  n is the total number of pieces of medical equipment.
  R_{rr} is the total number of repair requests.

  As per one study, the value of this index ranged from 0.3 to 2 [9].

- **Index II**

  This index measures how much time elapses from a customer request until the failed medical equipment is fully repaired and put back in service. The index is defined by

  $$\alpha_{at} = \frac{T_{tr}}{m} \tag{9.6}$$

  where

  α_{at} is the average turnaround time per repair.
  T_{tr} is the total turnaround time.
  m is the total number of work orders or repairs.

As per one study, the turnaround time per medical equipment repair ranged from 135 to 35.4 hours [9].

- **Index III**
  This index is a cost ratio and is defined by

$$\alpha_{cr} = \frac{C_{ms}}{C_{ma}} \tag{9.7}$$

where

α_{cr} is the cost ratio.
C_{ms} is the medical equipment service cost. It includes all parts, materials, and labor costs for unscheduled and scheduled service, including in-house, vendor, prepaid contracts, and maintenance insurance.
C_{ma} is the medical equipment acquisition cost.

For various classifications of medical equipment, a range of values for this index are available in Ref. [9].

9.6.2.2 Mathematical Models

Over the years, a large number of mathematical models have been developed for performing engineering equipment maintenance analysis. Some of these models can equally be used for performing medical equipment maintenance analysis. One of these models is presented below.

9.6.2.2.1 Model

This mathematical model can be used for determining the optimum time interval between item replacements. The model is based on the assumption that the item/equipment average annual cost is composed of average investment, operating, and maintenance costs. Thus, the average annual total cost of a piece of equipment is defined by

$$C_{at} = C_{of} + C_{mf} + \frac{C_{in}}{t_{ie}} + \frac{(t_{ie} - 1)}{2}[i + j] \tag{9.8}$$

where

C_{at} is the average annual total cost of a piece of equipment.
t_{ie} is the item/equipment life expressed in years.
C_{of} is the item/equipment operational cost for the first year.
C_{mf} is the item/equipment maintenance cost for the first year.
C_{in} is the investment cost.
i is the amount by which maintenance cost increases annually.
j is the amount by which operational cost increases annually.

By differentiating Equation (9.8) with respect to t_{ie} and then equating it to zero, we get

$$t_{ie}^* = \left[\frac{2C_{in}}{i+j}\right]^{1/2}$$

(9.9)

where

t_{ie}^* is the optimum time between item/equipment replacements.

Example 9.2

Assume that for a medical equipment, we have the following data values:

C_{in} = $200,000
j = $2,000
i = $500

Determine the optimum replacement period for the medical equipment under consideration.

By substituting the above specified data values into Equation (9.9), we obtain

$$t_{ie}^* = \left[\frac{2(200,000)}{500+2,000}\right]^{1/2}$$

$$= 40 \text{ years}$$

Thus, the optimum replacement period for the medical equipment under consideration is 40 years.

9.7   SOURCES FOR OBTAINING MEDICAL EQUIPMENT RELIABILITY-RELATED DATA

There are many organizations in the United States from which failure data directly or indirectly concerned with medical equipment can be obtained. Some of these organizations are as follows:

- Emergency Care Research Institute (ECRI), 5200 Butler Parkway, Plymouth Meeting, PA 19462, USA
- Center for Devices and Radiological Health (CDRH), Food and Drug Administration (FDA), 1390 Piccard Drive, Rockville, MD 20850, USA
- Parts Reliability Information Center (PRINCE) Reliability Office, George C. Marshall Space Flight Center, National Aeronautics and Space Administration (NASA), Huntsville, AL 35812, USA

- Government Industry Data Exchange Program (GIDEP) GIDEP Operations Center, Fleet Missile Systems, Analysis, and Evaluation Group, Department of Navy, Corona, CA 91720, USA
- National Technical Information Service, 5285 Port Royal Road, Springfield, VA 22161, USA
- Reliability Analysis Center (RAC), Rome Air Development Center (RADC), Griffis Air Force Base, Department of Defense, Rome, NY 13441, USA

Some of the data banks and documents considered quite useful for obtaining failure data concerning medical equipment are as follows:

- **Medical Device Reporting System (MDRS).** This system was developed by Center for Devices and Radiological Health (CDRH), Food and Drug Administration (FDA), 1390 Piccard Drive, Rockville, MD 20850, USA.
- **Universal Medical Device Registration and Regulatory Management System (UMDRMS).** This system was developed by Emergency Care Research Institute (ECRI), 5200 Butler Parkway, Plymouth Meeting, PA 19462, USA.
- **Hospital Equipment Control System (HECS).** This system was developed in 1985 by Emergency Care Research Institute (ECRI), 5200 Butler Parkway, Plymouth Meeting, PA 19462, USA.
- **NUREG/CR-1278.** Handbook of Human Reliability analysis with Emphasis on Nuclear Power Plant Applications, U.S. Nuclear Regulatory Commission, Washington, DC, USA.
- **MIL-HDBK-217.** Reliability Prediction of Electronic Equipment, Department of Defense, Washington, DC, USA.

9.8 USEFUL GUIDELINES FOR HEALTHCARE AND RELIABILITY PROFESSIONALS FOR IMPROVING MEDICAL EQUIPMENT RELIABILITY

There are a large number of professionals involved in the design, manufacture, and use of various types of medical equipment/devices. Reliability engineers and analysts are one of them. Nonetheless, some of the guidelines considered quite useful for reliability and other professionals for improving medical equipment reliability are as follows [26,42]:

- **Reliability Professionals**
  - Always focus on critical failures as not all device/equipment failures are equally important.
  - Keep in mind that manufacturers are totally responsible for reliability during the equipment/device design and manufacturing phase, and during its operational phase it is basically the responsibility of users.
  - Focus on cost effectiveness and always keep in mind that some reliability-related improvement decisions need very little or no additional expenditure.
  - Aim to use simple and straight forward reliability methods as much as possible instead of some sophisticated approaches used in the aerospace industrial sector.

- Use methods such as FMEA, qualitative FTA, design review, and parts review for obtaining immediate results.
- **Other professionals**
  - Keep in mind that the application of reliability principles have quite successfully improved the reliability of systems used in the aerospace sector, and their proper applications to medical equipment/devices can generate quite similar dividends.
  - Recognize that failures are the cause of poor medical equipment/device reliability, and positive thinking and measures can be quite useful for improving medical equipment/device reliability.
  - Compare human body and medical equipment/device failures. Both of them need appropriate measures from reliability professionals and doctors for enhancing equipment/device reliability and extending human life, respectively.
  - For the total success in regard to equipment/device reliability, both manufacturers and users must accept clearly their share of related responsibilities.
  - Remember that the cost of failures is probably the largest single expense in a business organization. These failures could be associated with equipment, business systems, people, etc., and reduction in these failures can reduce the cost of business quite significantly.

9.9   PROBLEMS

1. List at least six facts and figures concerned, directly or indirectly, with medical equipment/devices reliability.
2. Discuss the main classifications of electronic equipment/devices used in the health care system.
2. Compare FTA with FMEA with respect to medical equipment/devices.
4. Describe the parts count method.
5. List at least eight important operator errors that occur during medical equipment/device operation or maintenance.
6. List at least 15 medical devices with high incidence of human error.
7. Assume that the repair time of a medical equipment/system are exponentially distributed with a mean value (mean time to repair) of 8 hours. Calculate the probability that a repair will be completed in 20 hours.
8. Define medical equipment maintainability and list at least 8 frequently addressed maintainability design factors with respect to medical equipment.
9. Discuss classifications of medical equipment with respect to maintenance and repair.
10. List at least eight good sources for obtaining medical equipment/device reliability-related data.

REFERENCES

1. Hutt, P.B., A History of Government Regulation of Adulteration and Misbranding of Medical Devices, *The Medical Device Industry*, edited by N.F. Estrin, Marcel Dekker, Inc., New York, 1990, pp. 17–33.
2. Murray, K., Canada's medical device industry faces cost pressures, regulatory reform, *Medical Device and Diagnostic Industry Magazine*, Vol. 19, No. 8, 1997, pp. 30–39.
3. Crump, J.F., Safety and Reliability in Medical Electronics, *Proceedings of the Annual Symposium on Reliability*, 1969, pp. 320–330.
4. Johnson, J.P., Reliability of ECG instrumentation in a hospital, *Proceedings of the Annual Symposium on Reliability*, 1967, pp. 314–318.
5. Meyer, J.L., Some instrument induced errors in the electrocardiogram, *The Journal of the American Medical Association*, Vol. 201, 1967, pp. 351–358.
6. Gechman, R., Tiny flaws in medical design can kill, *Hospital Topics*, Vol. 46, 1968, pp. 23–24.
7. Taylor, E.F., The Effect of Medical Test Instrument Reliability on Patient Risks, *Proceedings of the Annual Symposium on Reliability*, 1969, pp. 328–330.
8. Dhillon, B.S., Bibliography of literature on medical reliability, *Microelectronics and Reliability*, Vol. 20, 1980, pp. 737–742.
9. Dhillon, B.S., *Medical Device Reliability and Associated Areas*, CRC Press, Boca Raton, Florida, 2000.
10. Allen, D., California home to almost one-fifth of U.S. medical device industry, *Medical Device and Diagnostic Industry Magazine*, Vol. 19, No. 10, 1997, pp. 64–67.
11. Medical Devices, Hearing Before the Subcommittee on Public Health and Environment, U.S. Congress House Interstate and Foreign Commerce, Serial No. 93-61, *U.S. Government Printing Office*, Washington, DC, 1973.
12. Banta, H.D., The regulation of medical devices, *Preventive Medicine*, Vol. 19, 1990, pp. 693–699.
13. Dhillon, B.S., *Reliability Technology in Health Care* Systems, *Proceedings of the IASTED International Symposium on Computers Advanced Technology in Medicine, Health Care, and Bioengineering*, 1990, pp. 84–87.
14. Schwartz, A.P., A Call for Real Added Value, *Medical Industry Executive*, February/March 1994, pp. 5–9.
15. Walter, C.W., Instrumentation Failure Fatalities, *Electronic News*, January 27, 1969.
16. Micco, L.A., Motivation for the Biomedical Instrument Manufacturer, *Proceedings of the Annual Reliability and Maintainability Symposium*, 1972, pp. 242–244.
17. Kohn, L.T., Corrigan, J.M., Donaldson, M.S., Editors, *To Err is Human: Building a Safer Health System*, Institute of Medicine Report, National Academy Press, Washington, DC, 1999.
18. MIL-HDBK-217, *Reliability Prediction of Electronic Equipment*, Department of Defense, Washington, DC.
19. RDH-376, *Reliability Design Handbook, Reliability Analysis Center*, Rome Air Development Center, Griffis Air Force Base, New York, 1976.
20. MIL-STD-1629, *Procedures for Performing a Failure Mode, Effects and Criticality Analysis*, Department of Defense, Washington, DC.
21. Palady, P., *Failure Modes and Effects Analysis*, PT Publications, West Palm Beach, Florida, 1995.
22. Rose, H.B., A Small Instrument Manufacturer's Experience with Medical Equipment Reliability, *Proceedings of the Annual Reliability and Maintainability Symposium*, 1972, pp. 251–254.

23. *Fault Tree Handbook, Report No. NUREG-0492*, U.S. Nuclear Regulatory Commission, Washington, DC., 1981.
24. Dhillon, B.S., Singh, C., *Engineering Reliability: New Techniques and Applications*, John Wiley, New York, 1981.
25. Singh, C., Reliability calculations on large systems, *Proceedings of the Annual Reliability and Maintainability Symposium*, 1975, pp. 188–193.
26. Dhillon, B.S., *Design Reliability: Fundamentals and Applications*, CRC Press, Boca Raton, Florida, 1999.
27. Shooman, M.L., *Probabilistic Reliability: An Engineering Approach*, McGraw Hill, New York, 1968.
28. Casey, S., *Set Phasers: and Other True Tales of Design Technology and Human Error*, Aegean, Inc., Santa Barbara, California, 1993.
29. Bogner, M.S., Medical Devices: A New Frontier for Human Factors, *CSERIAC Gateway*, Vol. 4, No. 1, 1993, pp. 12–14.
30. Sawyer, D., *Do It By Design: Introduction to Human Factors in Medical Devices*, Center for Devices and Radiological Health (CDRH), Food and Drug Administration, Washington, DC., 1996.
31. Novel, J.L., Medical device failures and adverse effects, *Pediatric Emergency Care*, Vol. 7, 1991, pp. 120–123.
32. Bogner, M.S., Medical Devices and Human Error in Human Performance, in *Automated Systems: Current Research and Trends*, Edited by M. Mouloua and R. Parasuraman, Lawrence Erlbaum Associates, Hillsdale, New Jersey, 1994, pp. 64–67.
33. Hayman, W.A., Human Factors in Medical Devices, in Encyclopaedia of Medical Devices and Instrumentation, edited by J.G. Webster, Vol. 3, John Wiley, New York, 1988, pp. 1542–1553.
34. Wikland, M.E., *Medical Device and Equipment Design*, Interpharm Press Inc., Buffalo Grove, Illinois, 1995.
35. Norman, J.C., Goodman, L., Acquaintance with and maintenance of biomedical instrumentation, J. Assoc. Advan. Med. Inst., Vol. 1, September 1966, pp. 8–10.
36. Waits, W., Planned Maintenance, *Medical Research Engineering,* Vol. 7, No. 12, 1968, pp. 15–18.
37. Grant-Ireson, W., Coombs, C.F., Moss, R.Y., Editors, *Handbook of Reliability Engineering and Management*, McGraw Hill, New York, 1988.
38. AMCP-133, *Engineering Design Handbook: Maintainability Engineering Theory and Practice*, Department of the Army, Washington, DC., 1976.
39. Blanchard, B.S., Verma, D., Peterson, E.L., *Maintainability*, John Wiley, New York, 1995.
40. Dhillon, B.S., *Engineering Maintainability*, Gulf Publishing, Houston, Texas, 1999.
41. Cohen, T., Validating Medical Equipment Repair and Maintenance Metrics: A Progress Report, *Biomedical Instrumentation and Technology*, Jan./Feb., 1997, pp. 23–32.
42. Taylor, E.F., The Reliability Engineer in the Health Care System, *Proceedings of the Annual Reliability and Maintainability Symposium*, 1972, pp. 245–248.

10 Mining Equipment Reliability

10.1 INTRODUCTION

Each year, a vast sum of money is spent to produce various types of equipment for use by the mining industry around the globe, and this expenditure is increasing quite rapidly. For example, in 2004, the United States mining equipment manufacturers shipped approximately $1.4 billion worth of goods, and a year later that figure jumped to around $2 billion [1]. Nowadays, the world economy is forcing mining companies to modernize their operations through increased automation and mechanization.

Thus, as equipment used in mines is becoming more complex and sophisticated, its cost is increasing rapidly. This in turn makes it quite cost-ineffective for having standby units. To meet production targets, mining companies around the globe are increasingly demanding better equipment reliability. Reliability is a very good indicator of overall equipment condition and is expressed as the probability that a piece of equipment will perform its function satisfactorily for the desired period of time when used as per stated conditions. However, it is to be noted that in the industrial sector reliability is often expressed in terms of mean time between failures.

This chapter presents various important aspects of mining equipment reliability.

10.2 REASONS FOR IMPROVING MINING EQUIPMENT RELIABILITY AND FACTORS IMPACTING MINING SYSTEM RELIABILITY

There are many reasons to improve reliability of mining equipment and some of these are as follows [2,3]:

- To lower the cost of poor reliability (the true poor reliability cost in most mining operations, when estimated effectively, is quite significant).
- To take advantage of lessons learned from other industrial sectors such as nuclear power generation, defense, and aerospace.
- To lower the performance of mining equipment-related services in an unplanned manner because of short notice.
- To maximize profit.

- To overcome challenges imposed by global competition.
- To provide more accurate short-term forecasts for operating hours of equipment.

There are many factors that, directly or indirectly, impact mining equipment/system reliability. Thirteen of these factors are as follows [2,3]:

- **Factor 1: Equipment failure.** It is basically a maintenance issue and it causes an interruption in the production process.
- **Factor 2: Refueling and lubrication**. In this case, stoppage of equipment for refueling and lubrication results in the interruption of the production process.
- **Factor 3**: **Minor production stoppages.** In this case, "comfort stops", such as minor adjustments, can interrupt the production process.
- **Factor 4: Routine maintenance.** In this case, routine servicing, overhauls, component replacements lead to interruptions in production while the equipment is taken out of service.
- **Factor 5: Accident damage.** In this case, it causes an interruption to the ongoing production process if the equipment has to be taken out of service for inspection or repair.
- **Factor 6: Spillage and housekeeping.** In this case, the need to stop and clean up spillage around the shovel or the dump area causes an interruption in the production process as well.
- **Factor 7: Shift changes and crib breaks.** In this case, every shift change and crib break usually leads to an interruption to the steady-state nature of the production operation.
- **Factor 8: The blast.** In this case, there is frequently a need to stop the equipment operation.
- **Factor 9: Ineffective blasting.** In this case, it can lead to problems, such as poor digging ability, in certain areas and unreliable equipment operation.
- **Factor 10: The mine plan.** In this case, it normally calls for equipment to be shifted on a periodic basis as different areas are to be mined.
- **Factor 11: Weather.** In this case, fog or rain can interrupt production.
- **Factor 12: Downstream process.** In this case, if a downstream process stops in a direct tipping situation, it can lead to an interruption in the mining operation.
- **Factor 13: Geology.** In this case, variability in digging conditions can result in the need for trucks or shovels to stop.

10.3 USEFUL RELIABILITY-RELATED MEASURES FOR MINING EQUIPMENT

There are many mining equipment, directly or indirectly, reliability-related measures. Five of these measures considered quite useful are presented below [2,3].

Measure I: Mean Time Between Failures

This is defined by

$$MTBF = \frac{(H_t - H_d - H_s)}{F} \qquad (10.1)$$

where

$MTBF$ is the mean time between failures.
F is the number of failures.
H_t is the total hours.
H_d is the downtime expressed in hours.
H_s is the standby hours.

Measure II: Production Efficiency

This measure may simply be described as the ratio of actual output from a piece of equipment/machine (which satisfies the required quality standards) to its rated output during the period it is operational. Nonetheless, production efficiency is defined by

$$E_p = \left[\left\{ \frac{P_a}{\left(H_t - H_d - H_s\right)} \right\} / C_r \right] (100) \tag{10.2}$$

where

E_p is the production efficiency.
P_a is the actual production.
C_r is the rated capacity expressed in units per hour.

Measure III: Utilization

This is expressed by

$$UT = \frac{\left(H_t - H_d - H_s\right)(100)}{\left(H_t - H_d\right)} \tag{10.3}$$

where

UT is the utilization.

Measure IV: Overall Equipment Effectiveness

This is defined by

$$E_{oe} = \left(AV_{me}\right)\left(E_p\right)\left(UT\right) \tag{10.4}$$

where

E_{oe} is the overall equipment effectiveness.
AV_{me} is the mining equipment/system availability.

Measure V: Availability

This is simply the proportion of time the equipment/system is able to be used for its intended purpose and is defined by

$$AV_{me} = \frac{(H_t - H_d)}{H_t}(100) \tag{10.5}$$

10.4   OPEN-PIT SYSTEM RELIABILITY ANALYSIS

Nowadays, the selection of equipment for modern open-pit mines has become a quite challenging issue in terms of availability, maintainability, reliability, productivity, etc. The overall system has grown to the level where the application of reliability principles has been considered to be very useful for satisfying the ever-growing technological requirements. The system has various types of loading and dumping machinery that, in turn, are arranged in different arrays. These arrays result in various types of sequencing systems. The malfunction of a single element in the sequences can cause part or total system failure.

Nonetheless, open-pit mines' each element (i.e., shovel, dumper, working face, dumping point, etc.) may be considered as an independent link to the open-pit mine chain system. The chain system's reliability analyses when it forms parallel and series networks are presented below [4].

10.4.1   OPEN-PIT PARALLEL SYSTEM

This type of network is formed when there is more than one unit of an open-pit system's components/units functioning simultaneously and at least one of these units/components must work normally for the system to succeed. For example, there are two shovels functioning simultaneously and at least one of the shovels must work normally for the system to succeed. In this case, both shovels form a parallel network and the network's reliability, if both shovels fail independently, using Chapter 3 and Dhillon [5], is expressed by

$$R_{pn} = 1 - (1 - R_1)(1 - R_2) \tag{10.6}$$

where

R_{pn} is the shovel parallel network reliability.
R_1 is the reliability of shovel number one.
R_2 is the reliability of shovel number two.

For constant failure rates of both the shovels, using Chapter 3 and Dhillon [5], we write

$$R_1(t) = e^{-\lambda_1 t} \tag{10.7}$$

$$R_2\left(t\right)=e^{-\lambda_2 t} \tag{10.8}$$

where

$R_1\left(t\right)$ is the reliability of shovel number one at time t.
λ_1 is the constant failure rate of shovel number one.
$R_2\left(t\right)$ is the reliability of shovel number two at time t.
λ_2 is the constant failure rate of shovel number two.

By substituting Equations (10.7) and (10.8) into Equation (10.6), we get

$$R_{pn}\left(t\right)=1-\left(1-e^{-\lambda_1 t}\right)\left(1-e^{-\lambda_2 t}\right)$$

$$=e^{-\lambda_1 t}+e^{-\lambda_2 t}-e^{-\left(\lambda_1+\lambda_2\right)t} \tag{10.9}$$

where

$R_{pn}\left(t\right)$ is the shovel parallel network reliability at time t.

By integrating Equation (10.9) over the time interval $[0,\infty]$, we get the following equation for the shovel parallel network mean time to failure:

$$MTTF_{pn}=\int_0^\infty R_{pn}\left(t\right)dt$$

$$=\frac{1}{\lambda_1}+\frac{1}{\lambda_2}-\frac{1}{\left(\lambda_1+\lambda_2\right)} \tag{10.10}$$

where

$MTTF_{pn}$ is the shovel parallel network mean time to failure.

Example 10.1

Assume that an open-pit system has two independent and non-identical shovels forming a parallel network (i.e., at least one shovel must operate normally for the system to succeed). The shovel number 1 and 2 constant failure rates are 0.004 failures per hour and 0.005 failures per hour, respectively.

Calculate the open-pit system reliability for a 100-hours mission and mean time to failure.

By substituting the given data values into Equation (10.9), we obtain

$$R_{pn}(100) = e^{-(0.004)(100)} + e^{-(0.005)(100)} - e^{-(0.004+0.005)(100)}$$

$$= 0.8702$$

Similarly, by substituting the given data values into Equation (10.10), we obtain

$$MTTF_{pn} = \frac{1}{0.004} + \frac{1}{0.005} - \frac{1}{(0.004+0.005)}$$

$$= 338.88 \text{ hours}$$

Thus, the open-pit system reliability and mean time to failure are 0.8702 and 338.88 hours, respectively.

10.4.2 Open-Pit Series System

In this case, the open-pit-mine components/units, i.e., shovel, dumper, dumping place, and working face, form a series network. This means all components/units must function normally for the system to succeed.

For independent components/units, the system reliability is expressed by [4,5]

$$R_{os} = R_{sh} R_{du} R_{dp} R_{wf} \tag{10.11}$$

where

R_{os} is the open-pit series system reliability.
R_{sh} is the shovel reliability.
R_{du} is the dumper or drum-truck reliability.
R_{dp} is the dumping-place reliability.
R_{wf} is the working-face reliability.

For constant failure rates of the shovel, dumper or dump truck, dumping place, and working face, using Chapter 3 and Dhillon [5], we get

$$R_{sh}(t) = e^{-\lambda_{sh}t} \tag{10.12}$$

$$R_{du}(t) = e^{-\lambda_{du}t} \tag{10.13}$$

$$R_{dp}(t) = e^{-\lambda_{dp}t} \tag{10.14}$$

$$R_{wf}(t) = e^{-\lambda_{wf}t} \tag{10.15}$$

where

$R_{sn}(t)$ is the shovel reliability at time t.
λ_{sh} is the shovel constant failure rate.

$R_{du}(t)$ is the dumper or dump-truck reliability at time t.
λ_{du} is the dumper or dump-truck constant failure rate.
$R_{dp}(t)$ is the dumping-place reliability at time t.
λ_{dp} is the dumping-place constant failure rate.
$R_{wf}(t)$ is the working-face reliability at time t.
λ_{wf} is the working-face constant failure rate.

By inserting Equations (10.12)–(10.15) into Equation (10.11), we obtain

$$R_{os}(t) = e^{-\lambda_{sh}t} e^{-\lambda_{du}t} e^{-\lambda_{dp}t} e^{-\lambda_{wf}t}$$

$$= e^{-\left(\lambda_{sh} + \lambda_{du} + \lambda_{dp} + \lambda_{wf}\right)t} \tag{10.16}$$

where

$R_{os}(t)$ is the open-pit series system reliability at time t.

By integrating Equation (10.16) over the time interval $[0, \infty]$, we get the following equation for the open-pit series system mean time to failure:

$$MTTF_{os} = \int_0^\infty R_{os}(t)\,dt$$

$$= \frac{1}{\left(\lambda_{sh} + \lambda_{du} + \lambda_{dp} + \lambda_{wf}\right)} \tag{10.17}$$

where

$MTTF_{os}$ is the open-pit series system mean time to failure.

Example 10.2

Assume that an open-pit system is composed of four components: shovel, dumper, dumping place, and working face and each component's reliability is 0.95, 0.98, 0.96, and 0.92, respectively. Calculate the open-pit system reliability if all its components fail independently and form a series network.

By substituting given data values into Equation (10.11), we obtain

$$R_{os} = (0.95)(0.98)(0.96)(0.92)$$
$$= 0.8222$$

Thus, the open-pit system reliability is 0.8222.

Example 10.3

Assume that in Example 10.2, the constant failure rates of shovel, dumper, dumping place, and working face are 0.009 failures per hour, 0.008 failures per hour, 0.007 failures per hour, and 0.006 failures per hour, respectively.

Calculate the open-pit series system reliability for a 100-hour mission and mean time to failure.

By substituting the given data values into Equation (10.16), we obtain

$$R_{os}(100) = e^{-(0.009+0.008+0.007+0.006)(100)}$$
$$= 0.0497$$

Similarly, by substituting the given data values into Equation (10.17), we obtain

$$MTTF_{os} = \frac{1}{(0.009+0.008+0.007+0.006)}$$
$$= 33.33 \text{ hours}$$

Thus, the open-pit series system reliability for a 100-hour mission and mean time to failure are 0.0497 and 33.33 hours, respectively.

10.5 DUMP-TRUCK TIRE RELIABILITY AND THE FACTORS AFFECTING THEIR LIFE

In opencast mines, the shovel-truck system is probably the most flexible system and its reliability and availability are the most important factors in successfully meeting their production target. Tires of dump trucks are an important element of the shovel-truck system and their reliability directly or indirectly affects opencast mines' overall production performance.

A study concerning dump-truck tires' reliability revealed that times to failure of the tires followed a normal distribution [6]. It simply means that the failure rate of dump-truck tires is not constant and the reliability of the tires has to be calculated by using the normal distribution representing the tire failure times.

There are many factors that can affect the life of dump-truck tire including over inflation, heat generation, speed and haul length, under-inflation, and tire bleeding [6,7]. Each of these five factors is described below.

- **Over inflation.** It reduces the amount of tread in contact with the ground and makes tires vulnerable to factors such as snags, impact fractures, and cuts.
- **Heat generation.** In this case, as rubber is a poor conductor, the heat generated through flexing will lead to heat accumulation. Here, it is to be noted that although the recommended load-speed inflation pressure ensures an equilibrium between heat generated and dissipated, any deviation can result in high pressure.

- **Speed and haul length.** In this case, speeds above 30 km per hour and a haul length more than 5 km can considerably affect the tires' life.
- **Under inflation.** It can lead to excessive flexing of the sidewalls and increase in internal tire temperature. In turn, this can cause permanent damage to tires such as casing breakup, radial cracks, and ply separation.
- **Tire bleeding.** In this case, lowering tire pressure after a long run is a normal practice. This causes premature failures, increases tire temperature, and makes an under inflated tire withstand high load (i.e., vehicle weight)

10.6 PROGRAMMABLE ELECTRONIC MINING SYSTEM FAILURES

Past experiences over the years indicate that various types of hazards can occur with programmable electronic mining system hardware or software failures [8]. Hardware failures are physical failures and are normally the result of wear and random events. They can involve any physical part/component of the system including power supplied, programmable electronic devices, sensors, and data communication paths. Random hardware failures include items such as mechanical defects, open circuits, short circuits, broken wires, corroded contacts, and dielectric failures.

In contrast, software failures take place due to systematic (functional) errors. Systematic errors include items such as operator errors, software bugs, design errors, requirement errors, management-of-change errors, and timing errors.

A study of data obtained from the US Mine Safety and Health Administration (MSHA), Queensland Mines in Australia, and New South Wales Mines in Australia, concerning programmable electronic-based mining systems, revealed that during the period 1995–2001 there were a total of 100 mishaps. The breakdown of these mishaps is shown in Figure 10.1 [7].

Both systematic failures and random hardware failures shown in Figure 10.1 are described below, separately.

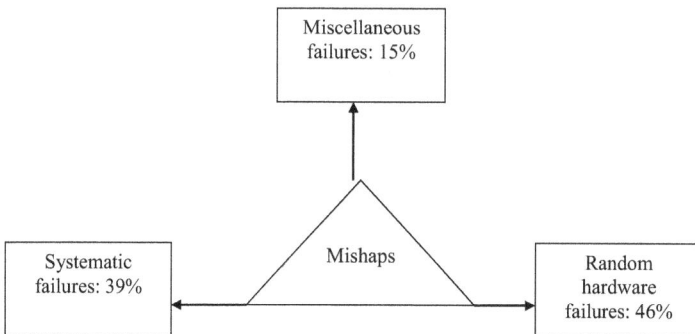

FIGURE 10.1 Mishaps, concerning programmable electronic-based mining systems, breakdown.

10.6.1 SYSTEMATIC FAILURES

Periodic systematic failures are also known as functional failures. Sources of these failures include hardware and software design errors, operator errors, errors made during maintenance and repair activities, and errors resulting from software modifications. For the period 1995–2001, analysis of systematic failures of the data for programmable electronic-based mining systems revealed the breakdown of the failures as follows [8]:

- Design-error-related failures: 50%
- Maintenance-and repair-error-related failures: 40%
- Miscellaneous failures: 10%

10.6.2 RANDOM HARDWARE FAILURES

The harsh environmental factors in mines, such as water and dirt intrusion, heat, shock, and vibrations, can significantly influence the occurrence of programmable electronic-based mining system hardware failures. These failures involve items such as electrical connectors, power supplies, sensors, solenoids, and wiring. An example of such failures is the degradation of rubber boots/seals used for keeping out dust and moisture.

For the period 1995–2001, analysis of random hardware failures of the data for programmable electronic-based mining systems revealed the following breakdown of the failures [8]:

- Sensor-related failures: 33%
- Electronic-component-related failures: 26%
- Moisture-related failures: 17%
- Actuator-related failures: 13%
- Miscellaneous failures: 11%

It is to be noted that in the above breakdown sensor-related failures include switch-related failures as well.

10.7   DESIGNING RELIABLE CONVEYOR BELT SYSTEMS AND METHODS OF MEASURING WINDER ROPE DEGRADATION

Nowadays, modern mining approaches are putting much greater emphasis than ever before on the reliability of the belt conveyor system that removes mined material, say coal, from the face. More specifically, with modern long-wall methods' application, it is quite possible that up to 90% of the mine production comes from one face, and it, in turn, must be handled effectively by only a single-gate conveyor. Thus, modern mines are looking for a conveyor reliability of around 100%, and the conveyor industrial sector is under increasing pressure for achieving this target.

This target or objective can be achieved by the industrial sector by the following five design-associated guidelines [9]:

- **Guideline I:** Design and plan for future requirements.
- **Guideline II:** Design for simplicity.
- **Guideline III:** Design for effective maintenance.
- **Guideline IV:** Design for unplanned events.
- **Guideline V:** Design for monitoring of equipment.

Winders that use steel wire ropes are generally used for moving materials in underground mines. During usage, the steel wire ropes are always subjected to continuous deterioration or degradation. Therefore, the ropes' reliability is very important for the performance of shaft-equipped mines as well as for the miners' safety. Past experiences, over the years, clearly indicate that the type of damage that occurs in winder ropes during their usage period includes, but not limited to, corrosion, abrasion formation of loops and kinks, and wire fatigue and resulting breaks [10–12]. Due to safety-related concerns, many mining regulatory authorities around the globe mandate periodic inspections to be conducted for determining the conditions of winder ropes. Thus, generally two types of inspections (i.e., magnetic non-destructive testing and visual inspection) are conducted as it is impossible to discover internal damage and corrosion through visual inspection alone.

Both magnetic non-destructive testing and visual inspection methods are described below, separately.

10.7.1   Magnetic Nondestructive Testing Method

Electromagnetic or permanent magnet-based methods are considered quite effective for detecting damage anywhere within a rope in both exterior and interior wires [11,13,14]. Magnetic rope testing is conducted by passing the rope through a permanent magnet-based device/instrument. In this case, rope's length is fully magnetized as it passes through the test device. Magnetic rope testing devices are utilized for monitoring two distinct types (i.e., types I and II) of magnetic field changes caused by the existence of anomalies.

Type I is called loss of metallic area (LMA) and it may simply be described as a relative measure of the amount of material mass missing from a cross-section in a wire rope. LMA is measured by comparing a section's magnetic field intensity with that of a reference section on the rope that represents the maximum, unworn metallic cross-section area. Type II changes involve a magnetic dipole produced by a discontinuity in a magnetized section of the rope such as a wire break, a groove worn into a wire, or a corrosion pit. It is to be noted that quite often these are known as leakage flux (LF) flaws.

10.7.2   Visual Inspection Method

This method is considered quite useful for highlighting changes in rope lay length and diameter, visible corrosion, crown wires' wear, external wire breaks or loose wires, and any other external damage. In fact, visual inspection is the only effective approach to highlighting the severity of rope external abrasive wear as the magnetic approach/method tends to underestimate the crown wire wear.

Seven steps of the visual inspection method are as follows [11,15]:

- **Step 1:** Measure the rope diameter and the lay length at a number of points/ sections.
- **Step 2:** Examine for broken wires as well as excessive crown wear.
- **Step 3:** Examine the entire rope end to end for damage or abuse.
- **Step 4:** Examine the rope termination for broken wires, condition of fastening, and corrosion.
- **Step 5:** Examine the sheaves for misfit, wear, etc.
- **Step 6:** Examine the drum's condition in the case of drum winders.
- **Step 7:** Check for appropriate lubrication.

10.8 TYPICAL MINING EQUIPMENT MAINTENANCE ERRORS AND FACTORS CONTRIBUTING TO MAINTENANCE ERRORS

There are many mining equipment-related maintenance errors that may, directly or indirectly, compromise safety. Some of the typical/common ones are as follows [3,16]:

- Failure to follow properly prescribed procedures and instructions
- Installation of wrong part
- Parts/components installed backward
- Failure to detect while inspecting
- Failure to check, calibrate, or align
- Use of wrong lubricants, greases, or fluids
- Omitting a part/component
- Failure to act on indicators of problems due to factors such as time constraints, priorities, or workload
- Error resulting from failure to complete task due to shift change
- Failure to lubricate
- Reassemble error
- Failure to close or seal properly

There are many factors that directly or indirectly contribute to mining equipment maintenance errors. Some of the important ones are as follows [3,15]:

- Lack of appropriate tools and troubleshooting guides
- Inaccessible parts/components
- Inability for making visual inspections
- Inappropriate placement of parts/components on equipment
- Excessive weight of parts being manually handled

- Poor manuals
- Poor provision for cable and hose management
- Poor layout of components/parts in a compartment
- Inadequate task inspection and checkout time
- Confined workspaces

10.9 USEFUL ENGINEERING DESIGN-RELATED IMPROVEMENT GUIDELINES FOR REDUCING MINING EQUIPMENT MAINTENANCE ERRORS

Over the years, engineering professionals working in the area of mining industrial sector have developed many useful engineering design-related improvement guidelines for reducing the occurrence of mining equipment maintenance-related errors. Six of the guidelines are as follows [16,17]:

- **Guideline I:** Improve equipment part interface by designing interfaces in such a way that the part can only be installed correctly and provide necessary mounting pins and other appropriate devices for supporting a part while it is being bolted or unbolted.
- **Guideline II:** Make use of decision guides for minimizing or reducing human guesswork by providing arrows for indicating correct hydraulic pressures, correct type of lubricants or fluids, and flow direction.
- **Guideline III:** Improve items, such as warning readouts, indictors, and devices, for reducing or minimizing human decision making.
- **Guideline IV:** Aim to improve fault isolation design by indicating the fault direction, designating appropriate test points and procedures, and providing built-in test capability.
- **Guideline V:** Design to facilitate detection of human errors.
- **Guideline VI:** Make use of operational interlocks in such a way that subsystems cannot be turned on when they are incorrectly installed or assembled.

10.10 PROBLEMS

1. What are the reasons for improving mining equipment reliability?
2. List at least 13 factors that impact mining equipment/system reliability.
3. What are the reliability-related measures for mining equipment? Define at least two such measures.
4. Assume that an open-pit system has two independent and identical shovels forming a parallel network. The shovel failure rate is 0.005 failures per hour.
   Calculate the open-pit system reliability for a 50-hours mission and mean time to failure.
5. Assume that an open-pit system is composed of four components: dumper, shovel, working face, and dumping phase; and failure probability of each component is 0.06, 0.05, 0.04, and 0.03, respectively.
   Calculate the open-pit system reliability if all its components fail independently and form a series network.

6. Assume that in question No. 5, the constant failure rates of dumper, shovel, working face, and dumping place are 0.004 failures per hour, 0.003 failures per hour, 0.002 failures per hour, and 0.001 failures per hour, respectively.

    Calculate the open-pit series system reliability for a 150-hour mission and mean time to failure.
7. What are the factors that can affect the life of dump-truck tires? Describe each of these factors.
8. Discuss programmable electronic mining system failures.
9. What are the typical/common mining equipment maintenance errors? List at least 12 of these errors.
10. Discuss useful engineering design-related improvement guidelines for reducing mining equipment maintenance errors.

REFERENCES

1. Chadwick, J., Higgins, S., US Technology, Int. Min., September 2006, pp. 44–54.
2. Dunn, S., Optimizing Production Scheduling for Maximum Plant Utilization and Minimum Downtime: The Reliability Revolution, Presented at the Dollar Driven Mining Conference, Perth, Australia, July 1997. Available online at http://www.plantmaintenance.com/ops-shtml.
3. Dhillon, B.S., *Mining Equipment Reliability, Maintainability, and Safety*, Springer-Verlag, London, 2008.
4. Mukhopadhyay, A.K., Open-pit system reliability, *Journal of Mines Metals and Fuels*, August 1988, pp. 389–392.
5. Dhillon, B.S., Design Reliability: Fundamentals and Applications, CRC Press, Boca Raton, Florida., 1999.
6. Dey, A., Battacharya, J., Banerjee, S., Prediction of field reliability for dumper tyres, *Int. J. Surf. Min. Reclamat. Environ.*, Vol. 8, 1994, pp. 23–25.
7. Balen, O., Off the road tyres: correct selection and proper maintenance, *Journal of Mines Metal Fuels*, Vol. 22, 1979, pp. 107–113.
8. Sammarco, J.J., Programmable Electronic Mining Systems: Best Practices Recommendations (in Nine Parts), Report No. IC 9480, (Part 6: 5.1 System Safety Guidance), National Institute for Occupational Safety and Health (NIOSH), US Department of Health and Human Services, Washington, DC, 2005. Available from the NIOSH: Publications Dissemination, 4676 Columbia Parkway, Cincinnati, OH 45226-1998.
9. Moody, C., Reliable conveyor belt design, *Proceedings of the American Mining Congress*, 1991, pp. 579–582.
10. Chaplin, C.R., Failure mechanisms in wire ropes, *Engineering Failure Analysis*, Vol. 2, 1995, pp. 45–57.
11. Kuruppu, M., Methods and reliability of measuring winder rope degradation, *Mine Planning and Equipment Selection*, April 2003, pp. 261–266.
12. Chaplin, C.R., Hoisting ropes for drum winders: the mechanics of degradation, *Mineral Technologies*, 1994, pp. 213–219.
13. Poffenroth, D.N., Procedures and Results of Electromagnetic Testing of Mine Hoist Ropes Using the LMA-TEST Instruments, Proceedings of the OIPEEC Round Table Conference, September, 1989, pp. 17–21.
14. Weischedal, H.R., The inspection of wire ropes in service: a critical review, *Material Evaluation*, Vol. 43, No. 13, 1985, pp. 1592–1605.

15. ASTM E157-93, *Standard Practice for Electromagnetic Examination of Ferro-Magnetic Steel Wire Ropes*, American Society for Testing and Materials (ASTM), Philadelphia, 1993.
16. Under, R.L., Conway, K., Impact of Maintainability Design on Injury Rates and Maintenance Costs for Underground Mining Equipment, in *Improving Safety at Small Underground Mines, Compiled by R.H. Peters, Special Publication No. 19–94*, Bureau of Mines, Department of the Interior, Washington, DC, 1994.
17. Dhillon, B.S., *Safety and Human Error in Engineering Systems*, CRC Press, Boca Raton, Florida, 2013.

11 Oil and Gas Industry Equipment Reliability

11.1 INTRODUCTION

Each year billions of dollars are spent around the globe to construct/manufacture, operate, and maintain various types of equipment/systems used in oil and gas industry. Nowadays, reliability of equipment used in the oil and gas industry has become an important issue due to various types of equipment reliability-related problems. For example, in 1996, corrosion-related failures including maintenance, annual direct cost in the US petroleum industry alone was $3.7 billion [1,2].

Today, the global economy is forcing all the companies involved with oil and gas to modernize their operations through increased automation and mechanization. Thus, as equipment used in the oil and gas industrial sector is becoming more complex and sophisticated, its cost is increasing quite rapidly. In order to meet production targets effectively, oil and gas companies are increasingly demanding better equipment reliability.

This chapter presents various important aspects of oil and gas industry equipment reliability.

11.2 OPTICAL CONNECTOR FAILURES

Optical fiber connectors are used for joining optical fibers in situations where a connect/disconnect capability is needed. In oil and gas applications, fiber-optic equipment including wet-mate optical connectors is an important part of the current subsea infrastructure. A study of reliability-related data (excluding cables or jumpers) collected over the period of 10-years reported four factors/issues (including their percentage breakdowns), as shown in Figure 11.1, that cause optical connector failures [3]. The data include field failures and the failures that took place during integration into equipment and testing process prior to field deployment.

It is to be noted from Figure 11.1 that 86% of the optical connector-related failures were due to material, external, and mechanical factors, and only 14% of failures were related to optical-related issues. Furthermore, when just field failures were studied, the optical connector-related failures occurred due to two types of issues only. These two types were mechanical issues and material issues. The mechanical issues accounted for 61% of the failures and the material issues accounted for 39% of the failures [3].

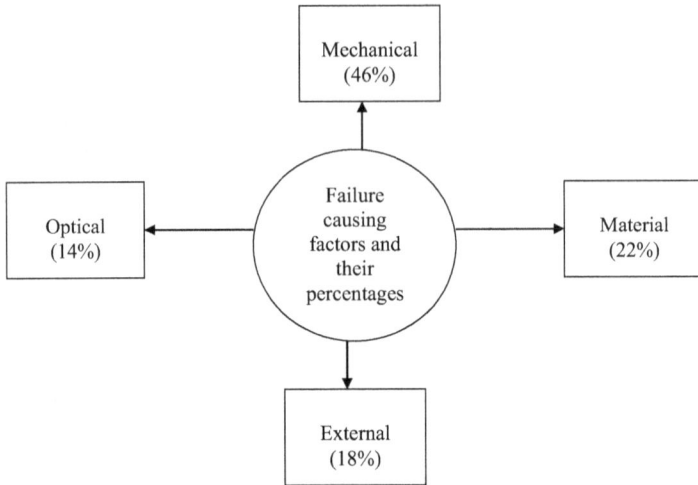

FIGURE 11.1 Factors along with their percentages in parentheses that caused optical connector failures.

Additional information on optical connector-related failures is available in Dhillon [3].

11.3 MECHANICAL SEALS' FAILURES

Mechanical seals have been increasingly used for many decades to seal rotating shafts. Nowadays, they are the most common types of seals found on items such as centrifugal pumps and compressors used in the oil and gas industrial sector. Over the years, mechanical seals' failure has become a very important issue. For example, a study carried out in a petroleum company reported that around 60% of plant breakdowns, directly or indirectly, were due to mechanical seal-related failures [4].

A study of mechanical seal failures conducted in a company reported the following ten causes for the failure of mechanical seals [4]:

- **Cause I:** External system or component failure (e.g., bearings)
- **Cause II:** Shaft and seal face plane misaligned
- **Cause III:** Auxiliary seal system failure (e.g., flush, cooling, recirculating, quench)
- **Cause IV:** Dry-running
- **Cause V:** Wrong seal spring compression
- **Cause VI:** Seal component failure (i.e., other than the faces or secondary seals)
- **Cause VII:** Highly worn carbon (i.e., greater than 4 mm of wear)
- **Cause VIII:** Hang-up (i.e., crystallization)
- **Cause IX:** Metal particles embedded in the carbon
- **Cause X:** Hang-up (i.e., coking)

11.3.1 MECHANICAL SEALS' TYPICAL FAILURE MODES AND THEIR CAUSES

Past experiences over the years indicate that mechanical seals can fail in many different failure modes due to various causes. Typical failure modes and their corresponding causes for mechanical seals are presented below [5].

- **Failure mode: Accelerated seal face wear.** Its causes are excessive torque, shaft-out-of-roundness, surface finish deterioration, misalignment, excessive shaft and play, inadequate lubrication, and contaminants.
- **Failure mode: Compression set and low-pressure leakage**. Its cause is extreme temperature operation.
- **Failure mode: Fractured spring.** Its causes are corrosion, stress concentration due to tooling marks, misalignment, and material flaws.
- **Failure mode: Seal embrittlement.** Its causes are idle periods between use, thermal degradation, fluid/seal incompatibility, and contaminants.
- **Failure mode: Torsional shear.** Its causes are excessive torque due to improper lubrication and excessive fluid pressure surges.
- **Failure mode: Excessive friction resulting in slow mechanical response**. Its causes are excessive seal swell, seal extrusion, excessive squeeze, metal-to-metal contact (i.e., out of alignment).
- **Failure mode: O-ring failure.** Its causes are installation error, excessive temperature (i.e., greater than 55°C), and excessive fluid pressure.
- **Failure mode: Open seal face-axial.** Its causes are impeller adjustment error, thrust movement, temperature growth, spiral failure (caused by conditions that allow some parts of the ring to slide and others to roll that cause twisting), etc.
- **Failure mode: Fluid seepage.** Its causes are insufficient seal squeeze (loss of spring tension) and foreign material on rubbing surface.
- **Failure mode: Seal fracture.** Its causes are excessive pressure velocity (PV) value, excessive fluid pressure on seal, and stress-corrosion cracking.
- **Failure mode: Seal face distortion.** Its causes are excessive PV value of seal operation, foreign material trapped between faces, insufficient seal lubrication, and excessive pressure on seal.
- **Failure mode: Clogged bellows.** Its causes are particles stuck at the inside of the bellows and hardening of fluid during downtime.
- **Failure mode: Axial shear.** Its cause is excessive pressure loading.
- **Failure mode: seal face edge chipping.** Its causes are excessive shaft deflection, seal faces out-of-square, and excessive shaft whip.
- **Failure mode: Open seal face-radial.** Its causes are shaft detection, shaft whip, and bent shaft.
- **Failure mode: Clogged spring.** Its cause is fluid contaminants.
- **Failure mode: Small leakage.** Its causes are insufficient squeeze and installation damage.

Additional information on mechanical seal failure modes is available in Skewis [5].

11.4 CORROSION-RELATED FAILURES

Corrosion in the oil and gas industrial sector has been acknowledged since the 1920s, and each year corrosion and other related failures still cost the oil and gas industrial sector hundreds of millions of dollars [6]. As per Kermani and Harrop [7], in the oil and gas industry corrosion-related failures constitute over 25% of failures.

A study carried out in the 1980s reported the following nine causes (along with their degree of contribution in percentages) for corrosion-related failure in petroleum-related industrial sector [7,8]:

- **Cause I:** CO_2 related: 28%
- **Cause II:** Preferential weld: 18%
- **Cause III:** H_2S-related: 18%
- **Cause IV:** Pitting: 12%
- **Cause V:** Erosion corrosion: 9%
- **Cause VI:** Galvanic: 6%
- **Cause VII:** Stress corrosion: 3%
- **Cause VIII:** Crevice: 3%
- **Cause IX:** Impingement: 3%

11.4.1 Types of Corrosion or Degradation that Can Cause Failure

The following ten possible types of corrosion or degradation that can cause failure [6]:

- **Type I: Galvanic corrosion.** It can take place in bimetallic connections at connections at opposite ends of the galvanic series that have enough potential for causing a corrosion reaction in the existence of an electrolyte.
- **Type II: Crevice corrosion.** This type of corrosion takes place in situations where crevice forms, such as partial penetration welds and backing strips, are employed.
- **Type III: Fretting corrosion.** It generally occurs in poorly lubricated valve stems where a partially opened valve causes some vibration that can result in galling and then, in turn, valve seizure and possible failure.
- **Type IV: Corrosion fatigue.** Over the years, it has played an important role in subsurface and drilling operations such as drill pipe and sucker-rod failures.
- **Type V: Impingement/cavitation.** Impingement can take place in situations where process fluid is forced to change its flow direction abruptly. Common offshore area for cavitation's occurrence is in pump impellors where pressure changes take place and high liquid flow rates occur.
- **Type VI: Erosion–corrosion.** It is observed quite often on the outer radius of pipe bends in oil and gas production due to rather quite high fluid flow rates as well as corrosive environments where flow exceeds 6 m/s for copper and nickel and 10 m/s for carbon steel.
- **Type VII: Microbiological-induced corrosion.** It is quite serious as it takes the form of localized pitting attack that can, directly or indirectly, cause a rapid loss of metal in a concentrated area, leading to leak or rupture.

- **Type VIII: Weight loss corrosion.** This type of corrosion occurs most commonly in the area of oil and gas production due to an electrochemical reaction between metal and the corrodents in the environment.
- **Type IX: Hydrogen-induced cracking.** In the past, this type of cracking has mostly took place in the controlled rolled pipeline steels and elongated stringers of non-metallic imperfections.
- **Type X: Stress corrosion cracking.** The most probable form of cracking phenomenon in the area of oil and gas production is sulfide and chloride stress corrosion cracking.

11.4.2 CORROSION/CONDITION MONITORING METHODS

Corrosion monitoring of internal surfaces may be conducted by using the combination of the following five methods [6]:

- **Method I:** Visual inspection
- **Method II:** Pipeline leak detection
- **Method III:** Intrusive probes and coupons for monitoring corrosion and erosion
- **Method IV:** Chemical analysis of samples taken from the product
- **Method V:** Measurements of non-intrusive wall thickness (radiography/ultrasonic)

The commonly used corrosion/condition monitoring methods are shown in Figure 11.2 [6].

Additional information on methods shown in Figure 11.2 is available in Price [6].

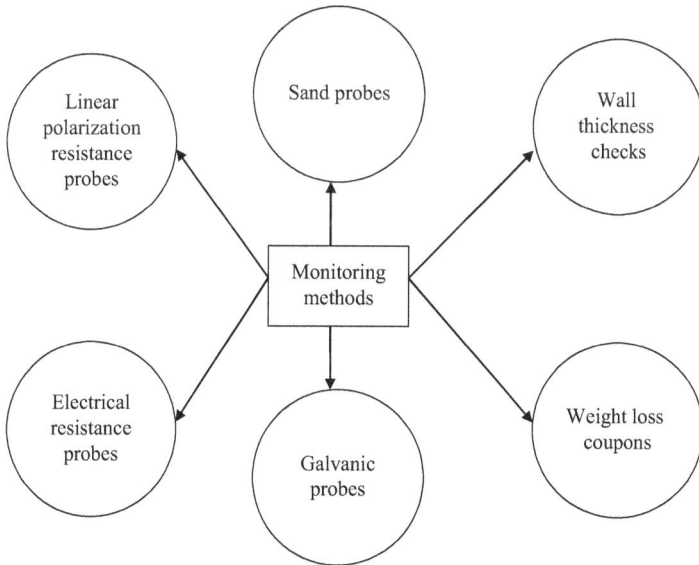

FIGURE 11.2 Commonly used corrosion/condition monitoring methods.

11.5 FATIGUE DAMAGE INITIATION ASSESSMENT IN OIL AND GAS STEEL PIPES

Nowadays, steel pipes are commonly used in offshore petroleum industrial sector. The fatigue behavior is a very important issue of concern in regard to the dynamic loadings (i.e., cyclic loadings). In oil and gas steel pipes, fatigue is one of the main failure causes observed, which in turn can result in catastrophic environmental damage as well as significant financial losses [9–11].

In order to assure oil and gas steel pipes' structural integrity and forewarn a fatigue-associated failure, it is very important to adopt a consistent fatigue criterion. Nonetheless, the fatigue associated damage may be divided into the following two main phases [9]:

- **Main phase I: Incubation phase.** During this phase, only microstructural-related changes, microcracking, and microcracks nucleation can be observed. It is to be noted that the study of this phase is more cumbersome to conduct since microstructural-related changes and fatigue damage cannot be easily separated.
- **Main phase II: Propagation phase.** During this phase, macrocrack propagation and macroscopic cracking result in fatigue failure, and the physical-related data that may quantify the material's damaged state can be more easily obtained.

Non-destructive evaluation (NDE) methods are considered quite useful for assessing fatigue life and evaluating structural integrity. These methods can assess fatigue damage's limitation, monitor changes in mechanical properties, and follow the fatigue damage process through the structures' life cycle subjected to cyclic loadings, in order to forewarn a malfunction. Five of these methods are as follows [9]:

- Magnetic evaluation method
- X-ray diffraction method
- Hardness measurements method
- Ultrasonics method
- Thermography method

It is to be noted that among the NDE methods for fatigue damage imitation assessments, X-ray diffraction method is generally considered to be one of the most suitable analysis methods [9]. Additional information on this method is available in Pinheira et al [9].

11.6 OIL AND GAS PIPELINE FAULT TREE ANALYSIS

The fault tree method was developed in the early 1960s at the Bell Telephone Laboratories for performing an analysis of the Minuteman Launch Control System in regard to safety [12]. Nowadays, the method is widely used around the world for conducting various types of reliability and safety studies. The method is described in detail in Chapter 4.

Here, this method's application for performing oil-gas long pipeline failure analysis is demonstrated through two examples presented below [13].

Example 11.1

Assume that an oil-gas pipeline failure can occur due to any of the following five events: misoperation, third-party damage, pipeline with defects, pipeline with serious corrosion, and material with poor mechanical properties. The occurrences of two of these events are described below:

- The event "Pipeline with serious corrosion" can occur due to the occurrence of the following events: Pipeline with poor corrosiveness resistance and corrosion. In turn, the event "corrosion" can be either due to internal corrosion or external corrosion.
- The event "Pipeline with defects" can be either due to pipeline with initial defects or pipeline with construction defects.

By using the fault tree symbols given in Chapter 4, develop a fault tree for the top event "oil-gas pipeline failure".

A fault tree for the example is shown in Figure 11.3. The single capital letters in Figure 11.3 denote corresponding fault events (e.g., A: Misoperation, B: Pipeline with defects, and C: Pipeline with serious corrosion).

Example 11.2

Assume that in Figure 11.3, the occurrence probabilities of independent fault events A, D, E, F, G, I, J, and K are 0.15, 0.14, 0.13, 0.12, 0.11, 0.10, 0.09, and 0.08, respectively. With the aid of Chapter 4, calculate the occurrence probability of the top event T: oil-gas pipeline failure.

The probability of the occurrence of event H is

$$P(H) = 1 - \left(1 - P(J)\right)\left(1 - P(K)\right) \tag{11.1}$$

$$= 1 - (1 - 0.09)(1 - 0.08))$$

$$= 0.1628$$

where

$P(J)$ is the probability of occurrence of event J.
$P(K)$ is the probability of occurrence of event K.

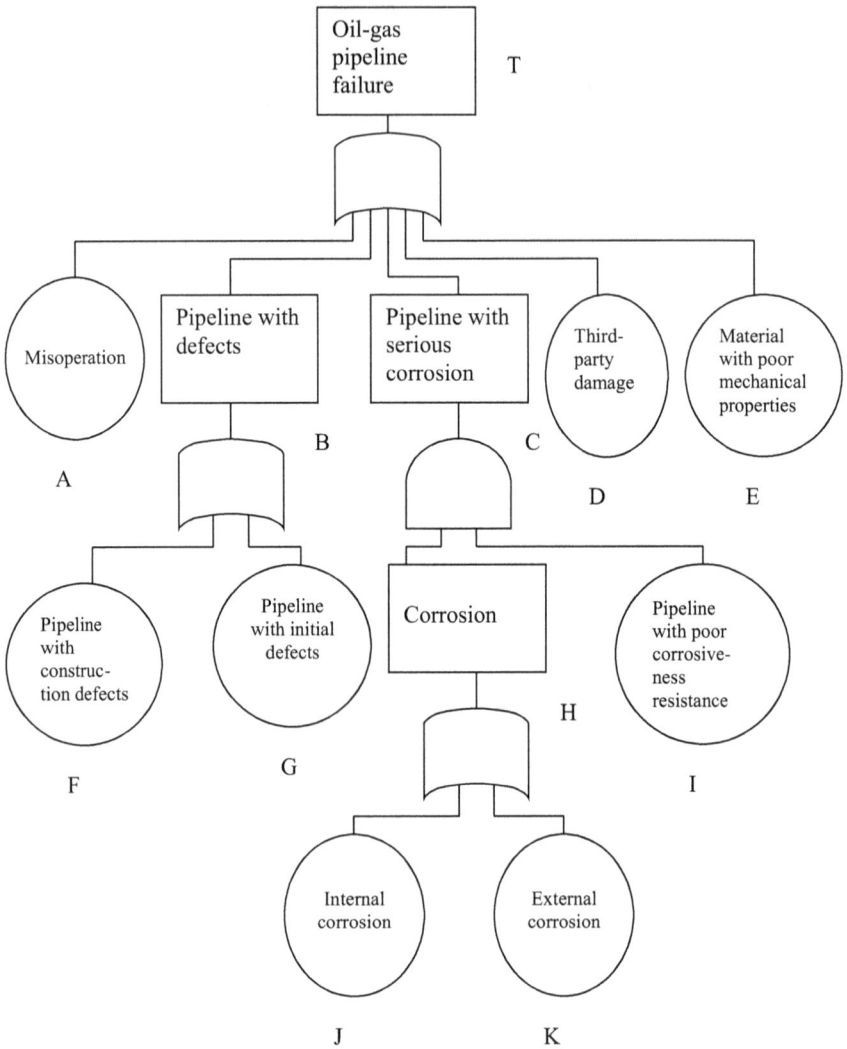

FIGURE 11.3   A fault tree for the top event: Oil-gas pipeline failure.

Similarly, the probability of the occurrence of event B is

$$P(B) = P(F)P(G) \qquad (11.2)$$
$$= 1 - (1 - 0.12)(1 - 0.11)$$
$$= 0.2168$$

where

$P(F)$ is the probability of occurrence of event F.
$P(G)$ is the probability of occurrence of event G.

The probability of the occurrence of the intermediate event C is

$$P(C) = P(H)P(I) \qquad (11.3)$$
$$= (0.1628)(0.10)$$
$$= 0.01628$$

The top event T (oil-gas pipeline failure) probability of occurrence is

$$P(T) = 1 - (1 - P(A))(1 - P(B))(1 - P(C))(1 - P(D))(1 - P(E))$$

$$= 1 - (1 - 0.15)(1 - 0.2168)(1 - 0.01628)(1 - 0.14)(1 - 0.13)$$

$$= 0.5100$$

Thus, the occurrence probability of the top event T (i.e., oil-gas pipeline failure) is 0.5100. Figure 11.3 fault tree with given and calculated fault event occurrence probability values is shown in Figure 11.4.

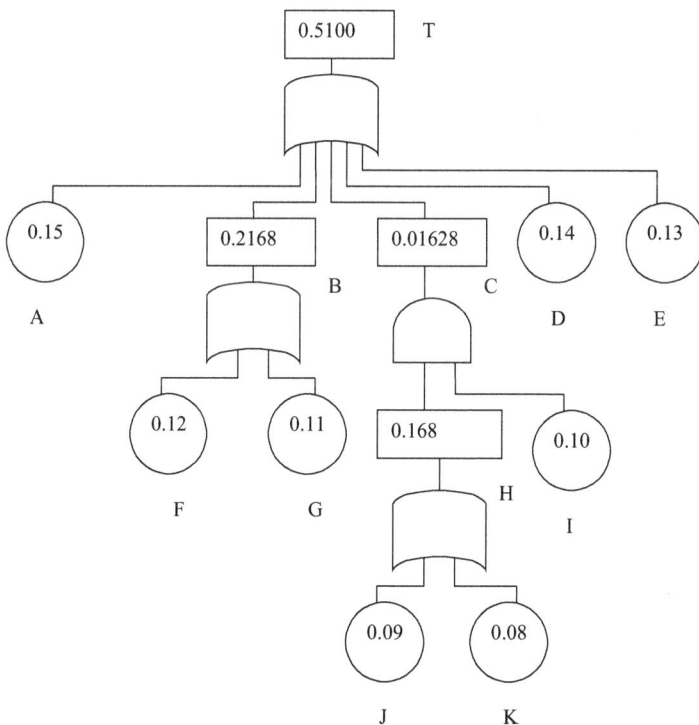

FIGURE 11.4 Redrawn Figure 11.3 fault tree with given and calculated fault event occurrence probability values.

11.7   COMMON CAUSE FAILURES DEFENSE APPROACH FOR OIL AND GAS INDUSTRY SAFETY INSTRUMENTED SYSTEMS

Safety instrumented systems (SIS) in the oil and gas industrial sector generally function in the low demand mode, which means that regular inspection and testing are absolutely necessary for revealing their failures. Past experiences over the years clearly indicate that common cause failures' occurrences are a serious threat to SIS reliability and may result in simultaneous failures of redundant parts/units and safety barriers [14–16].

Thus, a common cause failure may simply be expressed as any instance where multiple units/components/parts malfunction due to a single cause [16]. Some of the causes for common cause failures' occurrence are common manufacturer, design deficiency, common external power source, operation and maintenance errors, external normal environment, and external catastrophe.

The common cause failures defense approach described below for oil and gas industry SIS focuses on three key aspects presented below [15].

- To use the insight of failure causes for selecting efficient mechanism to defend against common cause failures' future occurrences.
- To highlight common cause failures and their associated causes on the basis of failure reports.
- To avoid introducing common cause failures during inspection and function testing-related processes.

11.7.1   COMMON CAUSE FAILURES DEFENSE APPROACH

The approach is based on six function testing and inspection tasks. These tasks are scheduling; preparation, execution, and restoration; failure reporting; failure analysis; implementation; and validation and continuous improvements. Thus, the approach is composed of six tasks based on checklists and analytical methods, such as operational sequence diagrams (OSD), influence diagrams, and cause-defense matrices. The six tasks are described below [15].

Task I: Scheduling

It is concerned with ensuring that during the scheduling process all appropriate improvements are captured. Here, it is to be noted that an important defense against common cause failures' occurrence is to ensure that any corrections and improvements to the test procedure are properly captured during the creation of a new function test or inspection work packages.

Task II: Preparation, Execution, and Restoration

It is concerned with avoiding the introduction of common cause failures during preparation, execution, and restoration. Three separate checklists presented below contain quite useful questions for preparation, execution, and restoration.

Preparation checklist questions:

- Have human error-associated incidents been experienced during earlier execution?
- Are all individuals involved with executing the test clearly familiar with the testing and calibration tools?
- Does the procedure describe all the essential steps for safely restoring the SIS?
- Have all types of potential human errors during execution and restoration been properly discussed and highlighted?
- Does the procedure contain all the known deficiencies (e.g., ambiguous instructions)?
- Have all compensating appropriate measures been clearly highlighted and implemented for avoiding human errors?
- Are all the calibration tools properly calibrated?

Execution checklist questions:

- Are all the parts operated within the specified operating and environmental conditions?
- Are all the parts appropriately protected against damage from nearby work-related activities?
- Are all the process connections free from plugging and (if applicable) heat-traced?
- Are all the field SIS parts (constituting the safety function under test) appropriately labeled?
- Are all the additional parts that are operated during SIS function testing and inspection process appropriately labeled?

Restoration checklist questions:

- Have all inhibits' and overrides' suspensions been appropriately verified and communicated?
- Has the safety function been appropriately verified prior to start-up?
- Are any remaining inhibits, bypasses, or overrides logged, and compensating measures appropriately highlighted and implemented?
- Has the physical restoration (e.g., bypasses and isolation valves) been appropriately verified?

Task III: Failure Reporting
It is concerned with improving failure reporting quality. In this regard, the following questions are considered very useful [15]:

- What appears to be the possible failure cause(s)?
- What was the effect of failure's occurrence on the overall safety function (i.e., loss of entire function, degraded, none at all)?

- Was the part tested or inspected in a different way than stated in the test or inspection procedure; if so, what was the reason for the approach to be different?
- How was the failure observed or found (i.e., during the repair or inspection process, by diagnostic, by review, upon demand, incidentally, or during function testing)?
- Have any similar types of failures occurred previously?
- Has the part/component been overexposed (i.e., environmental or by operational stresses); if so, what could be the associated causes?

Task IV: Failure Analysis

It is concerned with highlighting common cause failures through failure analysis. The following four steps are considered very useful in highlighting common cause failures [15]:

- **Step I:** Review failure's description and verify the initial failure classification (if necessary correct it).
- **Step II:** Conduct an appropriate initial screening that clearly captures failures that (i) share failure-associated causes, (ii) have been discovered within the framework of the same inspection or test interval, (iii) have quite similar design or physical/location, and (iv) the causes for failures are not random as stated by IEC 61508, 1998 and IEC 61511m 2003 documents [15].
- **Step III:** Conduct a root cause and coupling factors analysis by using influence diagrams.
- **Step IV:** List in a cause-defense matrix, all the root cause and coupling factors.

Task V: Implementation

It is concerned with implementing appropriate defensive measures. The proper implementation of common cause failures-associated defensive measures is very important for preventing the occurrences of similar types of failures. Additional information on this task is available in Lundteigen and Rausand [15].

Task VI: Validation and Continuous Improvements

It is concerned with validation and continuous improvements. In regard to validation, the following questionnaire considered useful [15]:

- Are all common cause failures systematically highlighted and analyzed, and proper defenses implemented for preventing their re-occurrences?
- Are all dangerous undetected failure modes clearly known and properly catered for in the function test and inspection-associated procedures?
- Are all safety function redundant channels appropriately covered by the function test or inspection-associated procedures?
- Are all requirements for the safety function appropriately covered by the inspection or function test-related procedure(s)?
- Are all disciplines concerned with SIS inspection, testing, maintenance, and follow-up properly familiar with the concept of common cause failures?

- Are all personnel using the calibration and test tools clearly familiar with their proper application?
- Are all failures detected upon real demands appropriately analyzed for verifying that they would have been detected during a function or inspection test?
- Are all changes in operating or environmental conditions properly captured and analyzed for essential modifications to the SIS or related procedures?
- Are all the test and calibration tools suitable and maintained according to the vendor recommendations?
- Are all the procedure-associated shortcomings appropriately communicated to the responsible personnel and followed up?
- Are all the diagnostic alarms appropriately followed up within the stated mean time to restoration?
- Are all the test-associated limitations (compared to the actual demand conditions) clearly known?
- Are all failures introduced during function testing and inspection processes captured, analyzed, and used for improving the associated procedures?

Finally, it is to be noted that for all the above questions, the answer "No" indicates a potential weakness in the defense against common cause failures' occurrence, and should be discussed for determining appropriate corrective measures.

11.8 PROBLEMS

1. Discuss optical connector failures.
2. List at least ten causes for the failure of mechanical seals.
3. Discuss at least 12 typical failure modes of mechanical seals.
4. Discuss at least nine types of corrosion or degradation that can cause failure.
5. List at least six commonly used corrosion/condition monitoring methods.
6. Discuss fatigue damage initiation assessment in oil and gas steel pipes.
7. Assume that in Figure 11.3, the occurrence probabilities of independent fault events A, D, E, F, G, I, J, and K are 0.04, 0.05, 0.06, 0.07, 0.08, 0.09, 0.10, and 0.11, respectively. Calculate the occurrence probability of the top event T (i.e., oil-gas pipeline failure) and also the reliability of the oil-gas pipeline.
8. Define a common cause failure. What are the causes of common cause failures' occurrence?
9. What are the causes for the occurrence of following four failure modes of mechanical seals:
   - Open seal face-axial
   - Accelerated seal face wear
   - Seal face distortion
   - Seal embitterment
10. Describe the common cause failures defense approach for oil and gas industry safety instrumented systems.

REFERENCES

1. Kane, R.D., Corrosion in Petroleum Refining and Petrochemical Operations, in *Metals Handbook*, Vol. 13C: Environments and Industries, edited by S.O. Cramer and B.S. Covino, ASM International, Metals Park, Ohio, 2003, pp. 967–1014.

2. Dhillon, B.S., *Safety and Reliability in the Oil and Gas Industry*, CRC Press, Boca Raton, Florida, 2016.

3. Jones, R.T., Thiraviam, A., *Reliability of Fiber Optic Connectors, Proceedings of the IEEE OCEANS Conference*, 2010, pp. 1–10.

4. Wilson, B., Mechanical Seals, *Industrial Lubrication and Tribology*, Vol. 47, No. 2, 1995, pp. 4.

5. Skewis, W.H., Mechanical Seal Failure Modes, Support Systems Technology Corporation, Gaithersburg, Maryland, retrieved on May 28, 2015 from website:http:docslide.us/documents/mechanical-seal-failure-modes.html.

6. Price, J.C., Fitness-for-purpose Failure and Corrosion Control Management in Offshore Oil and Gas Development, *Proceedings of the 11th International Offshore and Polar Engineering Conference,* 2001, pp. 234–241.

7. Kermani, M.B., Harrop, D., The Impact of Corrosion on the Oil and Gas Industry, *Society of Petroleum Engineers (SPE) Production and Facilities*, August 1996, pp. 186–190.

8. Kermani, M.B., Hydrogen Cracking and Its Mitigation in the Petroleum Industry, *Proceedings of the Conference on Hydrogen Transport and Cracking in Metals*, 1994, pp. 1–8.

9. Pinheira, B., et al., Assessment of Fatigue Damage Initiation in Oil and Gas Steel Pipes, *Proceedings of the ASME 30th International Conference on Ocean, Offshore, and Arctic Engineering*, 2011, pp. 1–10.

10. Lyons, D., *Western European Cross-Country Oil Pipelines 30-Year Performance Statistics*, Report No. 1/02, CONCAWE, Brussels, Belgium, 2002.

11. Paulson, K., *A Comparative Analysis of Pipeline Performance, 2000–2003*, National Energy Board, Alberta, Canada, 2005.

12. Dhillon, B.S., Singh, C., *Engineering Reliability: New Techniques and Applications*, John Wiley, New York, 1981.

13. Tian, H., et al., *Application of Fault Tree Analysis in the Reliability Analysis of Oil-gas Long Pipeline*, Proceedings of the International Conference on Pipelines and Trenchless Technology, 2013, pp. 1436–1446.

14. Summers, A.E., Raney, G., Common cause and common sense, designing failure out of your safety instrumented system (SIS), *ISA Transactions*, Vol. 38, 1999, pp. 291–299.

15. Lundteigen, M.A., Rausand, M., Common cause failures in safety instrumented systems on oil and gas installations: implementing defense measures through testing, *Journal of Loss Prevention in the Process Industries*, Vol. 20, 2007, pp. 218–229.

16. Dhillon, B.S., Proctor, C.L., Common-Mode Failure Analysis of Reliability Networks, *Proceedings of the Annual Reliability and Maintainability Symposium*, 1977, pp. 404–408.

Index

Absorption law 14
Advanced Research Project Agency network 99
Advisory Group on the Reliability of Electronic
    Equipment 1
Air Force Institute of Technology 1
Arithmetic mean 11–12
Associative law 14
Availability 186
Average service availability index 148
Aviation accidents 124–6

Bathtub hazard rate curve distribution 25
Bathtub hazard rate curve 31–2; burn-in
    period 31–2; useful-life period 31–2;
    wear-out period 31–2
Bedside television sets 168
Bell laboratories 99
Binomial distribution 21
Binomial method 71–3
Boole, G. 11, 13
Boolean algebra laws 13–14; absorption law 14;
    associative law 14; cummutative law 13;
    distributive law 14; idempotent law 14
Brake defects 123–4
Bridge network 46
British international helicopters Chinook
    accident 125
British Overseas Airways Corporation flight 781
    accident 126

Cardano, G. 11
Cardiac defibrillators 168
Cardiac pacemakers 168
Cause-defense matrices 208
Center for Devices and Radiological Health 172
Colorimeters 169
Common cause failure 208
Common cause failures defense approach 208–11
Communication network failures 100
Commuter rail service 126
Computation availability 102
Computation reliability 102
Corrosion/condition monitoring methods 203;
    electrical resistance probes 203; galvanic
    probes 203; linear polarization resistance
    probes 203; sand probes 203; wall thickness
    checks 203; weight loss coupons 203
Corrosion-related failures 202–3
Cummutative law 13

Customer average interruption frequency
    index 149

Decomposition approach 62–5
Definitions 14–20; cumulative distribution
    function 16; expected value 18; final-value
    theorem Laplace theorem 20; Laplace
    transform 19; probability density function 17;
    probability 14
Delta-star method 65–9
Diathermy equipment 169
Distributive law 14
Dump-truck tire reliability 190

Electric beds 168
Electric power 1
Electric robot 83–7
Electrical resistance probes 203
Electrocardiogram errors 167
Electrocardiograph recorders and monitors 169
Electrocardiographic monitors 168
Electroencephalograph recorders and
    monitors 169
Electronic tubes 1
Emergency Care Research Institute 167
Error cause removal program 76
Etruscans and Egyptians 167
Exponential distribution 22

Failure density function 32
Failure mode effects and criticality
    analysis 51
Failure modes and effect analysis 51–2
Fault masking 102–7; N-modular redundancy 107;
    triple modular redundancy 102–107
Fault tree analysis 52–57
Fermat, P. 11
Fiber-optic equipment 199
Food and Drug Administration 168
Forced derating 148
Forced outage hours 147
Forced outage rate 147
Forced outage 147

Galvanic probes 203
Gas analyzers 169
General approach 170–1
General reliability function 34
Government Industry Data Exchange Program 179

Hardness measurements method 204
Hardware failures 101
Hazard rate function 33
Hindu-Arabic numeral system 11
Hospital Equipment Control System 179
Hydraulic robot 87–90

Idempotent law 14
Influence diagrams 208
International Air Transportation Association 123
Internet outage catgories 114–15
Internet server system 116–18

Jelinski and Moranda model 110

k-out-of-m network 42

Laplace, P.S. 11
Linear polarization resistance probes 203
Los Angeles Airways flight 841 accident 126
Loss of load probability 150–1

Magnetic evaluation method 204
Magnetic nondestructive testing method 193
Maintainability design factors 173–4
Maintainability function 174–5
Maintenance errors 194
Malicious failures 101
Man-machine systems analysis 76
Markov method 57–61
Markov, A.A. 57
Mean deviation 11–13
Mean forced outage duration 148
Mean time between failures 184–5
Mean time to failure 35
Mean time to forced outage 148
Mean time to repair 173–4
Mean time to robot failure 79–81
Mean time to robot problems 82–3
Measuring service quality index 150
Mechanical seal failures 200–201
Mechanical seals 200
Medical Device Reporting System 179
Medical devices 168–9
Medical equipment maintainability 173–5
Medical equipment maintenance 175–7
Medical equipment/device operator errors 172
Microanalysis techniques for failure
    investigation 129–31; differential scanning
    calorimetry 130; Fourier transform infrared
    spectroscopy 131; thermogravimetric
    analysis 130; thermomechanical analysis
    130
Military and other reliability documents 6
Mills model 110–111

Mine Safety and Health Administration 191
Mining equipment maintenance errors 194–5
Mining equipment reliability measures 184–6;
    availability 186; mean time between
    failures 184–5; overall equipment effectiveness
    185; production efficiency 185; utilization 185
Minuteman launch control system 52
Musa model 111–113
Mysterious failures 100

National Symposium on Reliability and Quality
    Control 1
National Technical Information Service 179
Nelson model 110
Network reduction approach 61–2
New South Wales mines 191
N-modular redundancy 107
Nondestructive evaluation methods 204

Oil and gas steel pipes 204
Oil-gas pipeline failure 205
Open-pit parallel system 186–8
Open-pit series system 188–90
Operational sequence diagrams 208
Optical connector failures 199–200
Optimum time interval between item
    replacements 177–8
Overall equipment effectiveness 185

Pan Am flight 6 accident 126
Parallel network 39
Parts count method 169–70
Parts reliability information center 178
Peripheral device failures 100
Pinpoint method 115–16
Planet corporation 75
Power generator unit 151–8
Power transmission lines 158–63
Probability distributions 21–5; bathtub hazard rate
    curve distribution 25; binomial distribution 21;
    exponential distribution 22; Rayleigh
    distribution 23; Weibull distribution 24
Probability tree analysis 69–1
Production efficiency 185
Programmable electronic mining system
    failures 191–2

Quality circles 76
Queensland mines 191

Rail defects 126–7
Ramamoorthy and Bastani model 110
Random hardware failures 192
Rayleigh distribution 23
Reliability Analysis Center 179

Reliability networks 36–46; bridge network 46; k-out-of-m network 42; parallel network 39; series network 36; standby system 44
Reliability professionals 179
Respirators 168
Rim defects 124
Road and rail tanker failure modes 128
Robot downtime 95–6
Robot economic life 91
Robot hazard rate 81–2

Safety instrumented systems 208
Sand probes 203
Scheduled outage 147
Scythian 11
Series network 36
Ship-related failures 129
Shooman Model 110
Shovel-truck system 190
Software failures 101
Software metrics 108–9
Software reliability assessment methods 108–13
Software reliability models 109–13; Jelinski and Moranda model 110; Mills model 110–1; Musa model 111–3; Nelson model 110; Ramamoorthy and Bastani model 110
Schick and Wolverton Model 110
Shooman model 110
Sources for obtaining information on reliability 4

Spectrophotometers 169
Standby system 44
Steering system defects 124
System average interruption duration index 148
System average interruption frequency index 149
Systematic failures 192

Thermography method 204
Triple modular redundancy 102–7
Turkish Airlines flight 981 accident 125

Ultrasonic method 204
Ultrasound equipment 169
United Airlines flight 585 accident 125
United Airlines flight 859 accident 126
Universal Medical Device Registration and Regulatory Management System 179
US Air flight 427 accident 125
Utilization 185

Vehicle failures 123–4
Visual inspection method 193–4

Wall thickness checks 203
Weight loss coupons 203
Wheel chairs 168
Weibull distribution 24

X-ray diffraction method 204

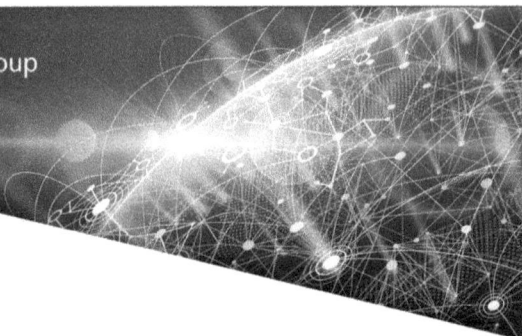

For Product Safety Concerns and Information please contact our EU
representative GPSR@taylorandfrancis.com
Taylor & Francis Verlag GmbH, Kaufingerstraße 24, 80331 München, Germany